Revise A2 Physics for OCR Specification A

David Sang

Heinemann is an imprint of Pearson Education Limited, a company incorporated in England and Wales, having its registered office at Edinburgh Gate, Harlow, Essex, CM20 2JE. Registered company number: 872828
Heinemann is a registered trademark of Pearson Education Limited

© David Sang, 2001

First published 2001

ISBN: 978 0 435583 42 2

08
10 9 8 7

Development editor Paddy Gannon

Edited by Patrick Bonham

Typeset and illustrated by Saxon Graphics Ltd, Derby

Index compiled by Ann Hall

Printed and bound in Great Britain by Ashford Colour Press Ltd, Gosport, Hampshire

Acknowledgements
The publishers have made every effort to trace the copyright holders, but if they have inadvertently overlooked any, they will be pleased to make the necessary arrangements at the first opportunity.

Tel: 01865 888058   www.heinemann.co.uk

Contents

Introduction – How to use this revision guide

This revision guide is for the OCR Physics A2 course (the second half of the A-level Physics course). It covers Module D of Specification A.

The module begins with an **introduction**, which summarises the content. It also reminds you of the topics from GCSE and from your AS course that the module draws on.

The content of the module is presented in **blocks**, to help you divide up your study into manageable chunks. Each block is dealt with in several spreads. These do the following:

- they **summarise** the content;
- they indicate **points to note**;
- they include **worked examples** of calculations;
- they include **diagrams** of the sort you might need to reproduce in tests;
- they provide **quick check** questions, to help you test your understanding.

At the end of the module, there are longer **end-of-module questions** similar in style to those you will encounter in tests. **Answers** to all questions are provided at the end of the book.

You need to understand the **scheme of assessment** for your course. This is summarised on pages 2–3 overleaf. At the end of the book, you will find a list of the various **formulae** and **definitions** you need to learn, and the others that are provided in tests.

A note about units

In the worked examples, we have included units throughout the calculations. (See, for example, the worked examples on pages 6 and 7.) This can help to ensure that you end up with the correct units in your final answer.

Scheme of Assessment for A-level Physics (OCR Specification A)

To follow the A2 course covered by this book, you need to have studied the AS course (though you may not yet have been assessed on the AS modules).

The AS course includes three **units of assessment** (A, B and C).

The A2 course also includes three units of assessment (D, E and F).

Units C and F include assessment of **experimental skills**, by coursework or practical examination.

The table opposite lists the units of assessment for the complete A-level course.

Notes on the written papers

The question papers for Units A, B, D and E have a common format, with two types of question.

- **Structured questions** require brief answers to several linked parts of a question.
- **Extended questions** require longer answers to a single question.
- Your answers to the extended questions will be used to assess the quality of your **written communication**.

Use the mark allocation and the space available for your answer to guide how much you write.

Synoptic assessment

In the A-level course, you are expected to develop your ability to make and use connections between different areas of physics. Also, you need to learn to apply ideas and skills developed through the course in contexts that may be new to you. These aspects of the course are assessed in Units E (options) and F1 (the synoptic paper).

- In the question papers for **Units E1–E5**, two questions will be set that require data analysis or comprehension. The context will relate to the option you have covered; they will draw on ideas from Units A–D.
- The **synoptic paper** (Unit F1) contains extended questions. These will draw on what you have learned in the different parts of the A-level course. You will be expected to show connections between different aspects of physics.

Practical examination

Your practical skills may be assessed by coursework or by practical examination. Each practical examination (C3 and F3) is in two parts:

A planning task Some days before the practical examination, you will be provided with a planning task. You may have access to laboratory facilities to carry out preliminary work; you may also wish to consult library books and other resources. Your written plan must be submitted by the deadline set by your examination centre. Your teacher will have to be sure that the work you submit is genuinely your own.

The practical examination You will be set a task that is different from the planning task, but in the same context. You must carry out the task, analyse the results, draw conclusions and evaluate the procedure. You may be given additional evidence to include in your analysis.

Make sure that you are familiar with the mark descriptors for each of the four practical skills.

Units of assessment – A-level Physics

Unit	Level	Name	Duration of written test	Types of question	Weighting at A-level
A	AS	Forces and motion	90 minutes	structured (75 marks); extended (15 marks)	15%
B	AS	Electrons and photons	90 minutes	structured (75 marks); extended (15 marks)	15%
C1	AS	Wave properties	60 minutes	structured (60 marks)	10%
C2*	AS	Experimental skills	none	coursework	10%
C3*	AS	Experimental skills	90 minutes	practical exam (60 marks)	10%
D	A2	Fields, forces and energy	90 minutes	structured (75 marks); extended (15 marks)	15%
E1–E5	A2	Options 1–5	90 minutes	structured (75 marks); extended (15 marks)	15%
F1	A2	Synoptic paper	75 minutes	extended (60 marks)	10%
F2*	A2	Experimental skills	none	coursework	10%
F3*	A2	Experimental skills	90 minutes	practical exam (60 marks)	10%

*C2 and C3 are alternatives; F2 and F3 are alternatives.

Notes on the units of assessment

- **Written tests** may be available in January and June.
- **Units E and F** must be taken at the end of the course, since these cover the synoptic aspect of the assessment.
- **Re-sits** are allowed once only; the better result counts, so you cannot end up with a lower score.
- Each unit of assessment has a **weighting** as indicated above; from this you can see that AS counts for 50% of A-level.
- **Aggregation** means combining the scores for each unit of assessment. You may enter for aggregation at the end of the AS course, or carry your marks forward to the A2 year. Since the rules for this are quite complex, you are strongly recommended to seek advice from your school's examinations officer.

A companion revision guide is available in this series for the AS part of the course.

Module D: Forces, fields and energy (A2)

This module has an equal weighting with Modules A and B. It extends several of the topics studied in Modules A–C of the AS course.

- **Blocks D1 and D2** consider forces and motion, and introduce the important concept of momentum. This makes it possible to analyse and solve many more problems. Block D2 considers circular motion and oscillations. Again, this relies on the application of your basic understanding of forces and motion.

- **Block D3** looks at force fields. This is a way of thinking about forces. How can one magnet attract another when they are not touching? How can the Sun pull on the Earth? Situations like this, where one object exerts a force on another 'at a distance', are described in terms of a field of force around the object. Gravitational, electrical and magnetic fields are used in a practical way to explain a variety of phenomena, including gravity, the operation of capacitors in electric circuits, and electromagnetic induction.

- **Block D4** returns to ideas about energy. In particular, it looks at how we can explain a variety of phenomena in terms of the energies (kinetic and potential) of the particles of which matter is made. It explains what is meant by temperature. It shows how the gas laws (relating pressure, volume and temperature of a gas) can be explained by considering the molecules of which the gas consists.

- Finally, **Block D5** looks at ideas about atoms and their underlying structure. The model of an atom consisting of a nucleus of protons and neutrons, with electrons orbiting the nucleus, allows us to explain radioactive decay. The energy released in radioactive decay can be calculated using the famous equation $E = mc^2$.

Two important pieces of maths appear in this module:

- the sine and cosine functions, used to represent the motion of an oscillating object;

- the exponential function, used to represent the decay of radioactive substances, and the discharge of a capacitor.

You will need to be able to use equations that include these functions.

Block D1: Dynamics, pages 6–13

Ideas from GCSE and Module A	Content outline of Block D1
<ul><li>Relationship between force, mass and acceleration</li><li>Force, work and distance</li><li>Kinetic and gravitational potential energy</li></ul>	<ul><li>Energy transfers and conversions</li><li>Conservation of energy</li><li>Momentum and its conservation</li><li>Collisions and explosions</li><li>Newton's laws of motion</li></ul>

Block D2: Oscillations and circular motion, pages 14–21

Ideas from GCSE and Module A	Content outline of Block D2
<ul><li>Displacement, velocity and acceleration–time graphs</li><li>Newton's laws of motion</li><li>Force, mass and acceleration</li><li>Kinetic and potential energy</li></ul>	<ul><li>Angles measured in radians</li><li>Centripetal acceleration and force</li><li>Simple harmonic motion</li><li>Resonance and damping</li></ul>

Block D3: Force fields, pages 22–35

Ideas from GCSE and Modules A and B	Content outline of Block D3
<ul><li>Acceleration caused by gravity</li><li>Static electricity</li><li>Force on a current-carrying conductor</li><li>Potential difference and e.m.f.</li></ul>	<ul><li>Gravitational fields</li><li>Electric fields and Coulomb's law</li><li>Capacitors and capacitance</li><li>Electromagnetic forces on currents and moving charges</li><li>Electromagnetic induction</li></ul>

Block D4: Thermal physics, pages 36–43

Ideas from GCSE and Modules A and B	Content outline of Block D4
<ul><li>States of matter, and changes of state</li><li>Energy transfers and work</li><li>Pressure in a fluid</li></ul>	<ul><li>Internal energy</li><li>Temperature scales</li><li>Specific heat capacity, specific latent heat</li><li>Gas laws and gas equations</li></ul>

Block D5: Nuclear physics, pages 44–53

Ideas from GCSE and Module C	Content outline of Block D5
<ul><li>Nature of α, β and γ radiations</li><li>Existence and structure of atoms</li><li>Background radiation</li><li>Diffraction of waves</li></ul>	<ul><li>Evidence for the structure of matter</li><li>Evidence for the structure of atoms</li><li>Mass–energy equivalence and $E = mc^2$</li><li>Nuclear fission, fusion and binding energy</li><li>Radioactive decay equations</li><li>Half-life and decay constant</li></ul>

End-of-module questions, pages 54–57

Module E: Options (A2)

One optional module must be selected from the following (these are not covered in this book). Each has an equal weighting with Modules A, B and D.

E1 Cosmology

E2 Health Physics

E3 Materials

E4 Nuclear and Particle Physics

E5 Telecommunications

Energy transfers

Energy can be transferred from one place to another; for example, electricity can transfer energy from a power station to your home. Energy can also be transferred from one object to another; if you push a car to get it moving, you are transferring energy to the car.

Here we consider mechanical **transfers of energy** (i.e. transfers involving forces). (Energy can also be transferred by heating and by electricity.)

Doing work

When a force moves through a distance in the direction of the force, it does **work** (see Module A, page 16 of the AS revision guide). The amount of **work done** tells us how much energy is being transferred by the force.

> **work done = force × distance moved *in the direction of the force* $W = F \times x$**

Units   If the force is in newtons (N) and the distance is in metres (m), the work done is in *joules* (J).

If the distance moved (displacement) is not in the same direction as the force, we have to calculate the *component* of the force in the direction of the displacement.

▶▶ *For a reminder of how to calculate the components of a vector, refer back to page 11 of the AS revision guide.*

✓ *Quick check 1*

Worked example

A crane lifts a 20 000 N load to the top of a tall building, as shown. Calculate the work done by the tension in the cable.

Step 1   Calculate the component of the tension in the direction of the displacement:

   component of T = $T \cos 40°$ = 20 000 N × cos 40° = 15 320 N

Step 2   Calculate the work done:

   work done = component of force × displacement
   = 15 320 N × 50 m = 766 kJ

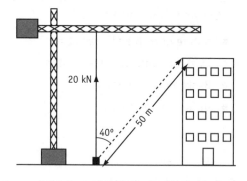

✓ *Quick check 2*

KE–GPE conversions

In many situations, an object's gravitational potential energy may be converted to kinetic energy, or vice versa. For example, when a ball rolls downhill, some of its GPE is converted to KE. Recall the following from Module A (pages 16–17):

> **kinetic energy $E_k = \frac{1}{2}mv^2$**
> **change in gravitational potential energy $\Delta E_p = mg\Delta h$**

Worked example

A ball initially moving at 5 m s^{-1} rolls up a smooth slope. How high up the slope will it rise? (Take g = 9.8 m s^{-2} and ignore the rotational KE of the ball.)

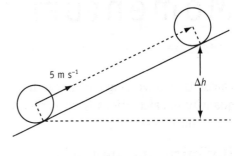

Step 1   The diagram shows the change in height Δh we have to calculate. Equate the ball's initial KE to its change in GPE:

$$\tfrac{1}{2}mv^2 = mg\Delta h$$

Step 2   Cancel m from both sides, rearrange, substitute values and solve:

$$\tfrac{1}{2}v^2 = g\Delta h$$

$$\Delta h = \tfrac{1}{2}v^2/g = \frac{\tfrac{1}{2}\times(5\ \text{m s}^{-1})^2}{9.8\ \text{m s}^{-2}} = 1.28\ \text{m}$$

Note that the ball's mass is irrelevant; it cancels out. A ball of any mass will reach the height calculated.

✔ *Quick check 3*

Conservation of energy

When energy is transferred from place to place, or converted from one form to another, the total amount always remains constant. This is the **principle of conservation of energy**. (We made use of this in the worked example above.)

However, during energy transfers and conversions, some of the energy may end up in a form that is not wanted. The energy transfer or conversion is said to be less than 100% efficient.

✔ *Quick check 4*

❓ Quick check questions

1   Are force and displacement scalar or vector quantities?

2   Look back at the worked example concerning the crane. Calculate the gain in height of the load as it is lifted to the top of the building, and hence its gain in GPE. Explain why this gives the same answer as the calculation in the worked example.

3   A stone is dropped from a height of 150 m. Calculate its speed when it reaches the ground. (Assume there is no air resistance, and take g = 9.8 m s^{-2})

4   Consider the falling stone in Question 3. In practice, it will be affected by air resistance. Use the ideas of work and conservation of energy to explain why the stone will be moving more slowly than you have calculated in Question 3.

Momentum

Calculations involving energy can help to solve many problems. Another useful quantity in calculations is *momentum*. The momentum of an object depends on its mass and its velocity.

Defining momentum

The **momentum** p of an object is the product of its mass m and its velocity v:

$$p = m \times v$$

An object has momentum *in a particular direction*. Hence momentum is a *vector* quantity. For example, the momentum of a woman of mass 60 kg running at 8 m s^{-1} due north is

$$p = m \times v = 60 \text{ kg} \times 8 \text{ m s}^{-1} = 480 \text{ kg m s}^{-1} \text{ due north}$$

The units of p are the units of m and v multiplied together: kg m s^{-1}. This can also be expressed as newton-seconds (N s). There is no special name for this unit.

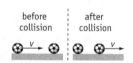

Conservation of momentum

Like energy, momentum is a quantity that is *conserved*; that is, for any event, the (vector) total amount of momentum before the event equals the (vector) total afterwards. This principle of conservation is made use of in solving problems – see pages 10 and 11. Here are some examples of situations to illustrate this idea. In each case, it is important to identify the **isolated system** (the system with no external forces) for which momentum is conserved.

- One ball rolls along and strikes a second, identical ball. The first stops dead; the second moves off with the speed of the first one. The momentum of the first ball has been transferred to the second. Isolated system = the two balls.

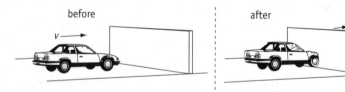

- The car runs into a wall and stops dead. Where has its momentum gone? It has been transferred to the wall, and hence to the Earth, which moves *very* slightly faster to the right! Isolated system = car + Earth.

- The ball falls towards the Earth, accelerating as it falls. It appears to be gaining momentum from nowhere. But the Earth is moving upwards, with a *very* small velocity. As the ball gains downward momentum, the Earth gains an equal and opposite upward momentum. The total momentum is zero. Isolated system = ball + Earth.

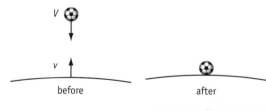

✔ *Quick check 3*

Momentum and kinetic energy

Both momentum (mv) and kinetic energy ($\frac{1}{2}mv^2$) depend on mass and velocity. This makes it very difficult to have separate mental images of these quantities. It is important to recall that:

- momentum is always conserved in an isolated system;
- kinetic energy is not always conserved.

Kinetic energy may be converted into other forms of energy, but momentum does not take different forms. A collision in which kinetic energy is conserved is described as **elastic**. Elastic collisions are springy – think back to the difference between elastic and plastic behaviour of materials: Module A, page 21.

▶▶ *More about collisions on pages 10–11.*

✓ *Quick check 4*

Momentum and force

It takes a *force* to change an object's momentum. The bigger the force, the faster the change in momentum. This equation *defines* what is meant by force:

> **force = rate of change of momentum**

(In calculus notation, $F = dp/dt$.)

This is an alternative way of stating Newton's second law of motion. It is a better way, because it can apply to an object whose mass, as well as its velocity, is changing.

✓ *Quick check 5*

? Quick check questions

1 Which of the following are *not* vector quantities: mass, velocity, momentum, kinetic energy?

2 A ship of mass 100 000 kg is sailing due west at 15 m s^{-1}. Calculate its momentum.

3 When a gun is fired, the bullet flies out very fast in a particular direction. The gun recoils more slowly in the opposite direction. Explain how momentum is conserved in this situation.

4 Which has more momentum, a boy of mass 40 kg running at 7.0 m s^{-1} or a girl of mass 32 kg running at 8.0 m s^{-1}? Which has more kinetic energy?

5 A stone of mass 0.4 kg is falling at 2.5 m s^{-1}. Half a second later, its downward velocity is 7.4 m s^{-1}. Calculate the stone's momentum at the start and finish of the 0.5 s interval. Use your answers to calculate the rate of change of its momentum, and hence the force acting on it.

Collisions and explosions

Collisions and explosions are examples of interactions between objects. In such situations momentum is conserved, and this is used to solve problems. In *elastic* collisions, kinetic energy is also conserved; this can help in calculations.

Collisions

We will consider only collisions in one dimension (i.e. along a line), but we could apply the same ideas to solve problems in two or three dimensions. Since momentum is conserved, we can write

momentum before collision = momentum after collision

▶ It is usually most helpful to start by drawing a pair of diagrams to show the situation before and after the interaction.

Worked examples

1 A trolley of mass 1 kg moving at 6 m s^{-1} collides with a second, stationary trolley of mass 2 kg. They stick together. With what velocity do they move off after the collision?

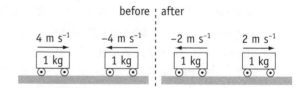

Step 1 Draw a before-and-after diagram; mark on it all the available information.

Step 2 Using 'momentum before collision = momentum after collision' and substituting values gives

$$\textbf{(1 kg} \times \textbf{6 m s}^{-1}\textbf{) + (2 kg} \times \textbf{0 m s}^{-1}\textbf{) = 3 kg} \times v$$

Step 3 Solve this equation for v:

$$\textbf{6 kg m s}^{-1} \textbf{= 3 kg} \times v$$

$$v = \textbf{2 m s}^{-1}$$

▶ You could omit 2 kg × 0 m s^{-1}, since this is obviously zero.

The velocity is in the direction the first trolley had been moving. It may be intuitively obvious that, since the mass increases by a factor of 3, the velocity decreases to a third of its initial value.

✓ *Quick check 1*

2 Two trolleys, each of mass 1 kg and moving at 4 m s^{-1}, collide head-on. They bounce apart; each has velocity 2 m s^{-1} after the collision. Show that momentum is conserved. Is kinetic energy also conserved in this collision?

Step 1 Draw a before-and-after diagram; mark on it all the available information. Give velocities to the right as positive, and to the left as negative.

Step 2 Here, we have to *show* that momentum is conserved. Calculate the momentum (mv) before and after collision *separately*:

momentum before = (1 kg × 4 m s^{-1}) + (1 kg × −4 m s^{-1}) = 0 kg m s^{-1}

momentum after = (1 kg × −2 m s^{-1}) + (1 kg × 2 m s^{-1}) = 0 kg m s^{-1}

Since momentum before collision = momentum after collision, we have shown that momentum is conserved.

Step 3 Calculate the kinetic energy ($\frac{1}{2}mv^2$) for each trolley, before and after the collision. Mark these values on the diagram.

kinetic energy before = $[\frac{1}{2} \times 1 \times 4^2] + [\frac{1}{2} \times 1 \times (-4)^2]$ = 8 J + 8 J = 16 J

kinetic energy after = $[\frac{1}{2} \times 1 \times (-2)^2] + [\frac{1}{2} \times 1 \times 2^2]$ = 2 J + 2 J = 4 J

So most of the kinetic energy disappears in the collision. (The trolleys may be deformed; heat and sound are produced.)

> ▶ Note that all values are positive, since squaring gets rid of the minus signs.

> ✓ *Quick check 2*

Explosions

Before an explosion, all parts of a system are at rest. Their combined momentum is zero. After the explosion, they are all flying apart. Each part has momentum, but their combined momentum *as a vector*, i.e. taking into account their different directions, is still zero.

Worked example

Two spring-loaded trolleys of masses 5 kg and 3 kg are stationary. When the spring is released, they fly apart. The lighter trolley moves at 4 m s^{-1}. How fast does the heavier one move?

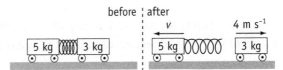

Step 1 Draw a before-and-after diagram; mark on it all the available information.

Step 2 The total momentum after the spring is released is zero, so:

momentum of heavier trolley + momentum of lighter trolley = 0

Step 3 Substitute values and solve for *v*:

$$(5 \text{ kg} \times v) + (3 \text{ kg} \times 4 \text{ m s}^{-1}) = 0$$

$$v = \frac{-12 \text{ kg m s}^{-1}}{5 \text{ kg}} = -2.4 \text{ m s}^{-1}$$

The minus sign means that the heavier trolley is moving in the opposite direction to the lighter one, i.e. to the left.

> ✓ *Quick check 3*

? *Quick check questions*

1 A car of mass 500 kg travelling at 24 m s^{-1} collides with a second, stationary car of mass 700 kg. The two cars move off together. What is their shared velocity?

2 A marble of mass 20 g is moving to the right at 4 m s^{-1} when it collides with a smaller, stationary marble of mass 8 g. The smaller marble moves off at 5 m s^{-1} to the right. With what velocity (magnitude and direction) does the first marble move after the collision?

3 A cannon of mass 400 kg fires a shell of mass 20 kg. If the shell leaves the cannon at 300 m s^{-1}, with what velocity does the cannon recoil?

Newton's laws of motion

This section of Module D has developed your understanding of forces and motion. You should now have a deeper understanding of Newton's laws of motion.

Newton's first law

> **An object continues in a state of rest or uniform motion in a straight line unless acted on by an unbalanced force.**

'Uniform motion' means constant velocity. An unbalanced force causes an object's velocity to change – it makes it *accelerate*.

✓ *Quick check 1*

Newton's second law

This law extends the first law to say what happens when an unbalanced force acts on an object. The object accelerates; the greater the force, the greater the acceleration.

> **When an unbalanced force acts on an object, it accelerates; its acceleration is proportional to the unbalanced force, and takes place in the direction of the force.**

You can think of this as $F \propto a$, or $F = ma$, where the object's mass m is the constant of proportionality. This tells us what is meant by *mass*:

> **The mass of an object is a measure of its resistance to change in its motion.**

Another word for mass in this sense is **inertia**. A better way to express Newton's second law is in terms of momentum (page 9):

> **When an unbalanced force acts on an object, its momentum changes; the rate of change of momentum is equal to the force producing it, and takes place in the direction of the force.**

This statement is a definition of what is meant by a *force* – something that causes a change in momentum. It is a more general statement of the second law, because it is possible that an object's mass may be changing, as well as its velocity.

✓ *Quick check 2*

Newton's third law

Two objects interacting with one another exert equal and opposite forces on each other, sometimes referred to as *action and reaction*. The two forces must:

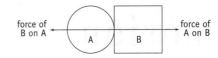

- be equal in magnitude but opposite in direction;
- be of the same type (e.g. both contact forces, or gravitational, etc.);
- act on different objects.

The third law applies for any two objects interacting with one another – they do not have to be in contact.

> **When two objects interact, the forces they exert on each other are equal and opposite.**

This is related to the conservation of momentum. Object A exerts force F on object B. This causes B's momentum to change. At the same time, B exerts force $-F$ on A, so A's momentum changes. The two changes in momentum are equal in magnitude but opposite in direction, because the forces are equal and opposite. If A's momentum increases by a certain amount, B's must decrease by an equal amount. Momentum is conserved.

✓ *Quick check 3*

The next block begins with circular motion, a topic in which the application of Newton's laws provides a proper understanding of how objects move along circular paths.

contact force of chair on person

weight of person

contact force of person on chair

? *Quick check questions*

1 If an object is in equilibrium, what can you say about the resultant force acting on it? What can you say about its velocity?

2 A rocket rises steadily upwards at a constant speed of 500 m s^{-1}. Its initial mass is 5×10^4 kg; after 10 minutes, this has decreased to 4.4×10^4 kg. Calculate the average force acting on the rocket during this time. This force does work on the rocket. Explain how this changes the rocket's energy.

3 The diagram shows a person sitting still on a chair. Three forces are shown. Which pair must be equal in magnitude because of Newton's third law? Which pair must be equal because of Newton's second law?

Describing circular motion

Many objects move along paths that are circular (or nearly circular) – a stone whirled around on the end of a piece of string, the Earth in its orbit around the Sun, a car along a curved stretch of road, an aircraft changing direction, an electron orbiting the nucleus of an atom. Since an object moving along a curved path is not moving in a straight line, it is an example of a moving object that is not in equilibrium.

Angular displacement; angles in radians

As an object moves along a circular path, it can be useful to state its position in terms of the angle θ through which it has moved relative to its starting position. This is called its **angular displacement** and is often given in **radians**, rather than degrees. The abbreviation for radians is **rad**.

start

- To convert from degrees to radians: multiply by $\pi/180$.
- To convert from radians to degrees: multiply by $180/\pi$.

$\theta = 1$ rad

Worked example

A car travels one-eighth of the way around a circular track. Through what angle θ has it moved in degrees, and in radians?

Step 1 Since a full circle is 360°, we can calculate the angle in degrees:

$$\theta = \frac{360°}{8} = 45°$$

2π rad = 360°

Step 2 Convert to radians:

$$\theta = 45 \times \frac{\pi}{180} \text{ rad} = \frac{\pi}{4} \text{ rad} = 0.79 \text{ rad}$$

π rad = 180°

$\frac{\pi}{2}$ rad = 90°

✓ *Quick check 1, 2*

Speed around a circular path

To calculate the speed of an object moving in a circular path, we need to know a distance and a time. For example:

$$\text{speed} = \frac{\text{circumference of circle}}{\text{time to complete one trip around circle}}$$

Since the circumference of a circle of radius r is $2\pi r$, if the time to complete one circuit is t we have

$$v = \frac{2\pi r}{t}$$

✓ *Quick check 3*

Constant speed, changing velocity

When an object moves in a circular path, its velocity is at a tangent to the circle. If it is moving at an unchanging speed, its *speed* is constant but its *velocity* is changing, because its direction of movement is changing.

The small arrow shows how the velocity vector changes from one position to the next. This arrow indicates the direction of the change in velocity, and hence the direction of the acceleration. It is directed towards the centre of the circle, and its magnitude is constant.

To produce this acceleration, there must be an unbalanced force of constant magnitude acting towards the centre of the circle. Hence the forces acting on an object with circular motion are unbalanced; it is not in equilibrium.

So, for uniform motion in a circle:

- the object moves at a constant speed;
- it has an acceleration towards the centre of the circle;
- this is caused by a force directed towards the centre of the circle.

The adjective that describes anything directed towards the centre of a circle is **centripetal**. Hence an object moving at a steady speed in a circular path has a centripetal acceleration caused by a centripetal force. The origins of centripetal forces are discussed on the next page.

✓ *Quick check 4*

? *Quick check questions*

1 Convert the following angles in degrees to radians: 360°; 180°; 90°; 60°; 45°.

2 Convert the following angles in radians to degrees: 1 rad; 0.25 rad; π rad; 2π rad; $\pi/5$ rad.

3 An aircraft is circling, waiting to land at an airport. Its circular path has a diameter of 20 km, and its speed is 120 m s^{-1}. How long will it take to complete one circuit of its path? In what time interval will the direction in which it is travelling change by 30°?

4 A toy train runs at a steady speed around a circular track. Which of the following are changing as it moves: its distance from the centre of the circle; its speed; its velocity; its centripetal acceleration; the centripetal force acting on it. Explain your answers.

Centripetal force and acceleration

A centripetal force is needed to keep an object moving along a circular path. Without such a force, the object would fly off in a straight line, at a tangent to the circle.

Take care! The word *centripetal* describes the direction of the force (towards the centre), but it doesn't tell you how the force arises. A centripetal force may arise in a variety of ways. Here are some examples.

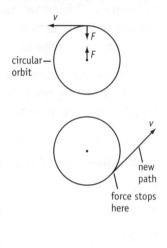

A stone is whirled around on the end of a string.	The tension in the string pulls the stone towards the centre of the circle.
A plane banks to follow a curved path.	The horizontal component of the lift force pushes the plane towards the centre of the circle.
A satellite orbits the Earth.	The Earth's gravitational pull on the satellite is directed towards the centre of the Earth.
An electron orbits the nucleus of an atom.	The electrostatic attraction of the nucleus pulls the electron towards it.

✓ *Quick check 1*

The size of the force

In each case, the moving object is pushed by the force towards the centre of the circle, but it never gets any closer. The force must be just large enough; any smaller, and the object will move away outwards; any bigger, and it will move inwards.

The centripetal force needed to make an object follow a curved path depends on three factors:

● the object's mass m: the greater the mass, the greater the force needed;

● the object's speed v: the greater the speed, the greater the force needed;

● the radius r of the path: the smaller the radius, the tighter the curve and the greater the force needed.

These quantities are combined in the following equation for F:

$$\text{centripetal force } F = \frac{mv^2}{r}$$

Worked example

A light aircraft of mass 500 kg is moving at a steady speed of 120 m s⁻¹ along a curved path of radius 2 km. What centripetal force is needed to keep it on this path?

Substitute values in the equation for F, and solve:

$$F = \frac{mv^2}{r} = \frac{500 \text{ kg} \times (120 \text{ m s}^{-1})^2}{2000 \text{ m}} = 3600 \text{ N}$$

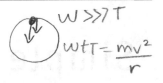

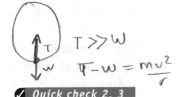

✓ *Quick check 2, 3*

Calculating centripetal acceleration

Since acceleration $a = F/m$, we have

$$a = \frac{v^2}{r}$$

(Strictly speaking, it is more correct to start from $a = v^2/r$ and use $F = ma$ to deduce that $F = mv^2/r$.)

Worked example

A spacecraft orbits above the Earth's surface at a height of 200 km. Its speed is 8 km s⁻¹. Calculate its centripetal acceleration. (Radius of Earth = 6400 km.)

Step 1 Calculate the radius of the orbit:

radius of orbit r = (6400 + 200) km = 6600 km

Step 2 Calculate the centripetal acceleration (note that kilometres must be converted to metres throughout):

$$\textbf{centripetal acceleration } a = \frac{v^2}{r} = \frac{(8000 \text{ m s}^{-1})^2}{6.6 \times 10^6 \text{ m}} = 9.7 \text{ m s}^{-2}$$

▶ Note that the answer is slightly less than the value of g at the Earth's surface, because the satellite is slightly further away from the centre of the Earth.

✓ *Quick check 4*

❓ *Quick check questions*

1 When a car follows a curved route, what force provides the necessary centripetal force?

2 A stone of mass 0.5 kg is whirled round at the end of a piece of string 40 cm in length. If it completes two complete revolutions in 1 s, what is the tension in the string?

3 A centripetal force is needed to make a car go round a bend. Use the equation $F = mv^2/r$ to explain why a bigger force is needed for a given speed when the car is following a more sharply curved bend.

4 The gravitational acceleration near the Moon's surface is 1.6 m s⁻². Calculate the speed of a satellite orbiting the Moon close to its surface. (Radius of orbit = 1800 km.)

▶ Think back to your study of the forces on vehicles in Module A.

▶ First calculate the stone's speed.

Simple harmonic motion

Start a pendulum swinging; pluck a stretched string; pull and release a mass on a spring. All of these result in **free oscillations**, in which a mass vibrates freely at its natural frequency. In many situations, these oscillations take the form of what is called **simple harmonic motion (SHM)**.

▶▶ *Forced oscillations are covered on page 21.*

Defining terms

An oscillation can be represented by a displacement–time graph, just like a wave. The following terms have the same meanings as for a wave:

- **displacement** x: distance of the mass from its equilibrium position (metres)

- **amplitude** A: greatest value of the displacement (metres)

- **period** T: time for one complete oscillation (seconds)

- **frequency** f: number of oscillations per second (hertz, Hz)

Frequency and period are related by

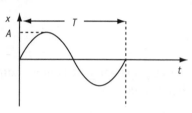

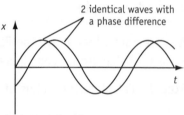

2 identical waves with a phase difference

$$f = \frac{1}{T} \quad \text{or} \quad T = \frac{1}{f}$$

We define a further quantity: **angular frequency** ω. We consider one complete oscillation or cycle as 2π radians. Then the angular frequency is the number of oscillations per second, measured in radians. It follows that

$$\omega = 2\pi f$$

Oscillations with the same frequency that reach their maximum displacements simultaneously are said to be **in phase** with one another. Oscillations with the same frequency that reach their maximum displacements at different times are said to be **out of phase** with one another. The **phase difference** between two such oscillations is given as a fraction of a cycle (2π radians).

$$\frac{1}{2} \text{ cycle phase difference} = \pi \text{ rad}$$
$$\frac{1}{4} \text{ cycle phase difference} = \pi/2 \text{ rad}$$

✓ *Quick check 1, 2*

Defining SHM

Not all oscillations are simple harmonic. For SHM, a mass is displaced from a central position, where it is in equilibrium. A restoring force F acts in the opposite direction to the displacement x; for SHM, this force must be proportional to x. This gives the mass an acceleration a, back towards the central position, that is proportional to x. Hence:

central position

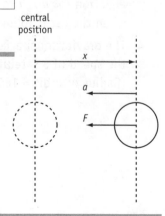

- Simple harmonic motion occurs when the acceleration of a mass is directed towards a fixed point and is proportional to its displacement from that point.

We can write this as an equation that involves the frequency f:

$$a = -\omega^2 x \quad \text{or} \quad a = -(2\pi f)^2 x$$

✓ *Quick check 3*

Back and forth

The mass speeds up as it approaches the midpoint. As soon as it passes through the midpoint, it starts to decelerate.

✓ *Quick check 4*

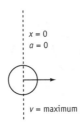

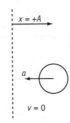

Here, the restoring force $F = ma$ has its greatest value. The mass is instantaneously at rest.
displacement $x = -A$
velocity $v = 0$
acceleration a = maximum

At the midpoint, the mass is moving fastest, although there is no force on it. It is in equilibrium ($F = ma = 0$).
displacement $x = 0$
velocity v = maximum
acceleration $a = 0$

Here, the mass's velocity is again zero, but the restoring force again has its maximum value.
displacement $x = +A$
velocity $v = 0$
acceleration a = maximum

Energy changes

The energy of the oscillating mass is transformed back and forth between kinetic and potential forms as it oscillates.

- Midpoint: maximum velocity, therefore maximum KE and zero GPE
- Endpoints: zero velocity, therefore zero KE and maximum GPE

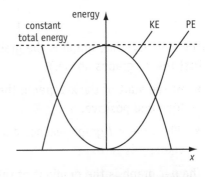

? *Quick check questions*

✓ *Quick check 5*

1 For the oscillation represented by the graph, what are the values of: amplitude, period, frequency, angular frequency?

2 The swings of a pendulum are timed. It completes 20 swings in 17.4 s. What are its period, frequency and angular frequency?

3 A mass is vibrating on the end of a spring. Its acceleration a (in m s^{-2}) is related to its displacement x (in m) from a fixed point by $a = -(40\pi)^2 x$. What are its angular frequency and its frequency? What is the period of its oscillation?

4 A pendulum swings from side to side. At what point in its oscillation is its speed greatest? At what point is its acceleration greatest?

5 Look at the energy graph above. When the mass is half-way between the midpoint and the endpoint of its oscillation, which is greater, its KE or its PE?

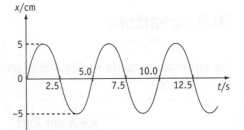

More about SHM

We can think of oscillations that are simple harmonic as being 'pure' oscillations. They give an x–t graph that is a sine curve.

SHM graphs

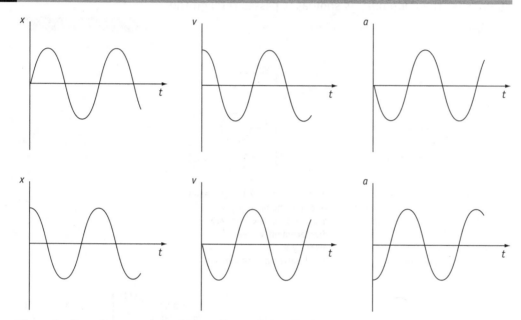

The velocity–time graph is the gradient of the displacement–time graph. (See the first row of graphs above.)

- At the start of the x–t graph, the gradient is steep and positive, so the velocity is high and positive.
- When the x–t graph reaches a peak (a maximum), its gradient is zero. The velocity is zero.

The a–t graph is the gradient of the v–t graph. It has troughs where the x–t graph has peaks.

> ▶ Recall that the acceleration is in the opposite direction to the displacement.

> ✓ *Quick check 1*

SHM equations

The information contained in a displacement–time graph can also be represented by an equation:

$$x = A \sin 2\pi ft \quad \text{or} \quad x = A \cos 2\pi ft$$

The difference between these equations is this:

- If the oscillation starts ($t = 0$) at the midpoint ($x = 0$), use the sine version (because sin 0 = 0).
- If the oscillation starts ($t = 0$) at the endpoint ($x = A$), use the cosine version (because cos 0 = 1).

Take care! When using these equations, ensure that your calculator is working in radians, not degrees!

> ✓ *Quick check 2, 3*

Damping

If the oscillating mass loses no energy, it will oscillate for ever with the same amplitude. However, if it loses energy, we say the oscillations are **damped**. Their amplitude decreases.

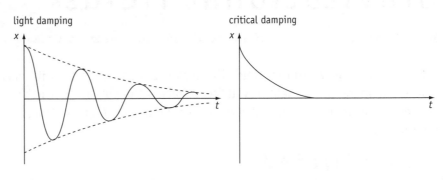

- Light damping: the amplitude decreases gradually as the mass oscillates.

- Critical damping: the damping is heavy enough for the displacement just to decrease to zero (the equilibrium position) without oscillation. With a little less damping, the mass overshoots the midpoint.

Damping is caused by frictional forces, e.g. drag of the air, or viscous drag in oil. A car suspension system is usually critically damped, so that the passengers do not bounce up and down each time the car passes over a bump in the road.

Resonance

It may be possible to force a mass to oscillate at any frequency. This is called a **forced oscillation**. If the forcing frequency matches the natural frequency of free oscillations, the amplitude increases to a large value. This is **resonance**.

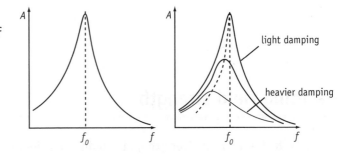

The graph shows that:

- at frequencies slightly above and below resonance, the amplitude is less;

- damping reduces the amplitude at resonance, and tends to shift the resonant frequency.

Examples of resonance are pushing a child on a swing (each push increases the amplitude slightly) and tuning a radio (the resonant frequency of the tuning circuit is adjusted to match the frequency of the signal).

✓ *Quick check 4*

? *Quick check questions*

1 The graph shows one oscillation for a vibrating mass. Copy the graph, and beneath it sketch the corresponding velocity–time and acceleration–time graphs.

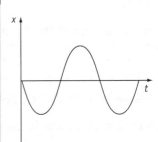

2 A mass oscillates such that its displacement x in cm is represented by the equation

$$x = 4 \cos (0.6t)$$

What are the values of the amplitude and frequency of this motion? What is the value of x when $t = 5$ s?

3 Write an x–t equation for oscillations of amplitude 0.2 m and frequency 0.5 Hz, for a mass whose initial displacement is zero. Calculate its displacement when $t = 0.2$ s.

4 Look at the resonance graph above. What happens to the sharpness of the resonance curve as damping increases?

Gravitational fields

The Earth has a **gravitational field**. This means that, if an object with mass is placed anywhere in that field, it will feel a force – the pull of the Earth's gravity. This force has another name – the **weight** of the object. A gravitational field is a field of force.

Representing a field

Field lines (lines of force) represent a gravitational field.

- The arrows show the direction of the force on a mass placed in the field.
- Lines closer together represent a stronger field.

Near the Earth's surface, the field is uniform. The field lines are effectively parallel; the force on an object is the same at all positions in the field.

On a larger scale, the Earth has a spherical field. The field lines diverge; the field gets weaker the further we move away from the surface.

For a uniform sphere, the external field is the same as if all of its mass were concentrated at the centre.

✓ Quick check 1

Defining field strength

The **gravitational field strength** g at a point in a field is the force per unit mass that acts on an object placed at that point.

Since force per unit mass is F/m, we can write

$$g = \frac{F}{m} \quad \text{or} \quad F = mg$$

You should recognise this as the equation used to calculate the weight of an object of mass m.

On the surface of the Earth, g has the approximate value

$$g = 9.8 \text{ N kg}^{-1}$$

This varies only slightly over the surface of the Earth.

Note that this value is the same as that of the acceleration of free fall, 9.8 m s^{-2}.

▶ The force that makes an object fall is gravity. Comparing $F = mg$ with $F = ma$ shows that $a = g$.

✓ Quick check 2

Newton's law of gravitation

Newton's law tells us how to calculate the gravitational force F between two objects of masses m_1 and m_2 separated by a distance r :

$$F = \frac{Gm_1 m_2}{r^2}$$

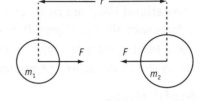

Here $G = 6.67 \times 10^{-11}$ N m^2 kg^{-2} is the **universal gravitational constant.**

This is an example of an *inverse square law*: F is proportional to $1/r^2$.

The objects are *point masses*, as if all of an object's mass is concentrated at its centre of gravity. Each of the two objects feels the same force (even if their masses are different), but in opposite directions. They are an equal and opposite pair of forces, as described by Newton's third law of motion (page 13).

▶▶ *Centre of gravity – see Module A, page 19.*

▶▶ *For another inverse square law, see page 26.*

g for a point mass

For a point mass m, the gravitational field strength g at a distance r is

$$g = \frac{Gm}{r^2}$$

Worked example

Find the mass of the Earth, given that $g = 9.8$ N kg^{-1} at its surface. (Radius of the Earth $= 6.4 \times 10^6$ m, $G = 6.67 \times 10^{-11}$ N m^2 kg^{-2}.)

Step 1   Rearrange the equation $g = Gm/r^2$:

$$m = \frac{gr^2}{G}$$

Step 2   Substitute values for g, r and G, and calculate the result:

$$m = \frac{9.8 \text{ N kg}^{-1} \times (6.4 \times 10^6 \text{ m})^2}{6.67 \times 10^{-11} \text{ N m}^2 \text{ kg}^{-2}} = 6.0 \times 10^{24} \text{ kg}$$

✓ *Quick check 3*

❓ *Quick check questions*

1 Draw diagrams including field lines to explain the following. When you go upstairs, your weight is effectively unchanged; if you climb Mount Everest, your weight decreases very slightly; if you are in a spacecraft 200 km above the Earth's surface, your weight is significantly less than on the surface.

2 The Moon's gravitational field strength is 1.6 N kg^{-1}. Calculate the weight of a 5 kg rock on the Moon. If dropped, how far will it fall in 1 s?

3 Two asteroids, of masses 4×10^{10} kg and 8×10^{10} kg, are separated by 20 km in space. Calculate the gravitational force each exerts on the other. Draw a diagram to show the directions of these forces.

Electric fields

Gravitational fields are created by objects with mass. **Electric fields** are created by objects with electric charge. One important difference is that there is only one type of mass, but there are two types of charge, *positive* and *negative*. Therefore electrostatic forces can be *attractive* or *repulsive*, whereas gravitational forces are always attractive.

Field lines

An electric field can be represented by **field lines**, rather like a gravitational field.

- The arrows show the direction of the force on a *positive* charge placed in the field.
- Arrows come out of positive charges, and go into negative charges.
- Lines closer together represent a stronger field.

Isolated positive charge.

Isolated negative charge.

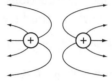

Like charges repel.

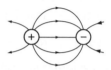

Unlike charges attract.

Outside an isolated charged, conducting sphere, the field is the same as that of a point charge at its centre.

Between two charged, parallel plates, the field is uniform.

With more charge, the field is stronger.

With plates further apart, the field is weaker.

✓ *Quick check 1*

Electric field strength

An electric field is a field of force. Any charged object placed in the field will feel a force. To define the **electric field strength** at a point in the field, we picture placing a positive charge Q at the point. Measure the force F that acts on the charge. Then

electric field strength = force per unit charge $E = F/Q$

The unit of electric charge is the *coulomb* (C). The unit of E is therefore *newtons per coulomb* ($N\ C^{-1}$).

▶▶ *Electric charge and the coulomb – see Module B, page 28.*

Recall from Module B, page 34, that one volt is the work done (in joules) in pushing one coulomb of charge round a complete circuit: $1\ V = 1\ J\ C^{-1}$. Also, from Module A, page 16, $1\ J = 1\ N \times 1\ m$. Therefore:

$$1\ N\ C^{-1} = 1\ J\ m^{-1}\ C^{-1} = 1\ V\ m^{-1}$$

So electric field strength can also be expressed in *volts per metre*.

✓ *Quick check 2*

Charged parallel plates

There is a uniform electric field between a pair of parallel plates. The plates can be charged by applying a voltage across them. The greater the voltage, the stronger the field.

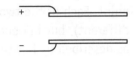

$$\text{strength of field } E = \frac{V}{d}$$

This is easily recalled if you remember that electric field strength can be measured in volts per metre (see above).

✓ *Quick check 3–5*

Now we have two equations for E:

$$E = \frac{F}{Q} \quad \text{and} \quad E = \frac{V}{d}$$

Putting these equal gives $F/Q = V/d$, and hence

$$Fd = QV$$

Thus the work done (Fd) in moving a charge Q through p.d. V is equal to QV.

This is the equation that defines the volt: $W = QV$ (see Module B, page 34).

❓ *Quick check questions*

1 Copy the diagram, which shows two charged, parallel metal plates. Add field lines to show the electric field between the plates. Explain how your diagram shows that this field is uniform.

2 An electron is moving through an electric field of strength 10 kN C^{-1}. What is the electric force on it? (Charge on an electron $e = 1.6 \times 10^{-19}$ C.)

3 Two parallel plates separated by 20 cm are connected to the terminals of a 60 V power supply. Calculate the electric field strength in the space between them.

4 Use the equation $E = V/d$ to explain the following:

 • increasing the p.d. between a pair of parallel plates increases the field strength between them;

 • increasing the separation of the plates decreases the field strength between them.

5 A dust particle carrying a charge of 4 mC is in the space between two parallel plates separated by 5 cm. If the plates are charged to 24 V, calculate the electric force on the dust particle.

Coulomb's law

The electric field around a charged sphere or point charge is *radial*, like the gravitational field around a spherical mass. **Coulomb's law** tells us how to calculate the force between two spherical charges.

Force between two spherical charges

The electrostatic force F between two charges Q_1 and Q_2 separated by a distance r is given by Coulomb's law:

$$F = \frac{kQ_1Q_2}{r^2}$$

The value of the constant k is approximately 9×10^9 N m^2 C^{-2}. It is usually written as $k = 1/(4\pi\varepsilon_0)$, where ε_0 is called the **permittivity of free space**.

The charges are *point charges*, as if all of an object's charge is concentrated at a point. Each of the two charges feels the same force (even if their charges are different), but in opposite directions. They are an equal and opposite pair of forces, as described by Newton's third law of motion (page 13).

> Like Newton's law of gravitation (page 23), this is an example of an *inverse square law*: F is proportional to $1/r^2$.

✓ *Quick check 1*

Worked example

In an oxygen atom, the outermost electron orbits the nucleus at a distance of approximately 0.09 nm. The electron charge is $-e$ and the nuclear charge is $+8e$, where $e = 1.6 \times 10^{-19}$ C. Calculate the electrostatic force exerted by the nucleus on the electron.

Step 1 Write down the known values of quantities:

$$k = 9 \times 10^9 \text{ N m}^2 \text{ C}^{-2}$$
$$Q_1 = e = 1.6 \times 10^{-19} \text{ C}$$
$$Q_2 = 8e = 8 \times 1.6 \times 10^{-19} \text{ C} = 1.28 \times 10^{-18} \text{ C}$$
$$r = 0.09 \text{ nm} = 0.09 \times 10^{-9} \text{ m}$$

Step 2 Write down the equation for Coulomb's law. Substitute values and calculate F:

$$F = \frac{kQ_1Q_2}{r^2}$$
$$= \frac{9\times10^9 \text{ N m}^2 \text{ C}^{-2} \times 1.6\times10^{-19} \text{ C} \times 1.28\times10^{-18} \text{ C}}{(0.09\times10^{-9} \text{ m})^2}$$
$$= 2.3 \times 10^{-7} \text{ N}$$

> There is no need to move the decimal point for r; your calculator will cope.

> Note how the units cancel correctly to leave N.

✓ *Quick check 2*

E for a point charge

For a point charge Q, the electric field strength E at a distance r is

$$E = \frac{kQ}{r^2}$$

This comes from dividing the Coulomb's law equation by Q_2. Since there is only one charge, we don't need to call it Q_1.

This equation corresponds to $g = Gm/r^2$ for gravitational field strength (see page 23).

✓ *Quick check 3*

Comparing gravitational and electric fields

	Gravitational field	Electric field
Origin	any object with mass	any object with charge
Field strength equation	$g = F/m$	$E = F/Q$
Radial field	point mass or spherical mass	point charge or charged sphere
Inverse square law	Newton's law: $F = \dfrac{Gm_1 m_2}{r^2}$	Coulomb's law: $F = \dfrac{kQ_1 Q_2}{r^2}$
Constant	$G = 6.67 \times 10^{-11}$ N m^2 kg^{-2}	$k = 9 \times 10^{9}$ N m^2 C^{-2}
Uniform field	near surface of Earth $g \sim 9.8$ N kg^{-1}	between charged parallel plates $E = V/d$

The equation $g = F/m$ can be applied in *any* gravitational field, and $E = F/Q$ in *any* electric field. The other equations can be applied only in radial or uniform fields, as appropriate.

? *Quick check questions*

1 Draw a diagram to show two negative point charges. Add arrows to represent the force each charge exerts on the other. What can you say about the magnitudes and directions of these forces?

2 Calculate the electric force between two protons in the nucleus of an atom, separated by 10^{-15} m. (Proton charge = $+e$ = $+1.6 \times 10^{-19}$ C.)

3 A metal sphere of radius 5 cm carries a charge of 40 mC. Calculate the electric field strength at a distance of 1 cm from the surface of the sphere. ($k = 1/4\pi\varepsilon_0 = 9 \times 10^{9}$ N m^2 C^{-2}.)

First calculate the distance from the centre of the sphere.

Capacitors

Capacitors are components used in circuits to store electric charge. They usually consist of two parallel metal plates, separated by a narrow gap.

Stored charge

When a capacitor is connected to a source of voltage (p.d.), one plate gains positive charge $+Q$, while the other gains negative charge $-Q$. The capacitor is said to be storing Q coulombs of charge.

Increasing the voltage pushes more charge onto the capacitor. The greater the voltage V and the greater the capacitance C of the capacitor, the more charge Q it stores. This is represented by the equation

> **charge stored = capacitance × p.d.** $Q = CV$

Definitions

The equation $Q = CV$ can be rearranged to define capacitance:

$$C = \frac{Q}{V}$$

The capacitance of a capacitor is the charge stored for each volt of p.d. across it.

Units   Capacitance is measured in **farads** (F). One farad is one coulomb per volt. Most practical capacitors have values measured in smaller units (see Appendix 4):

- 1 μF = 1 microfarad = 10^{-6} F
- 1 pF = 1 picofarad = 10^{-12} F

✓ *Quick check 1*

Capacitors in series and in parallel

Note that these are similar to the formulae for combining resistances (Module B, pages 30–31) but the other way round, with the reciprocal formula ($1/C$) for capacitances in series but for resistors in parallel.

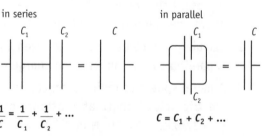

in series

$$\frac{1}{C} = \frac{1}{C_1} + \frac{1}{C_2} + \ldots$$

In series: combined capacitance is less than each individual capacitance.

in parallel

$$C = C_1 + C_2 + \ldots$$

In parallel: combined capacitance is the sum of the individual capacitances.

Worked example

What is the capacitance of a 20 µF and a 30 µF capacitor connected in series?

Step 1 Write down the appropriate formula, substitute values and add the fractions:

$$\frac{1}{C} = \frac{1}{C_1} + \frac{1}{C_2} = \frac{1}{20 \ \mu F} + \frac{1}{30 \ \mu F} = \frac{5}{60 \ \mu F} = \frac{1}{12 \ \mu F}$$

Step 2 Take the reciprocal to find C:

$$C = 12 \ \mu F$$

▶ Note that we can work in µF, without writing the powers of 10.

▶ If you prefer not to add fractions, use your calculator. This will give $1/C = 0.0833 \ \mu F^{-1}$ and $C = (1/0.0833) \ \mu F = 12 \ \mu F$.

✓ *Quick check 2, 3*

Storing energy

When a capacitor is charged up, work is done by the p.d. that pushes the charge onto the plates. This means that a charged capacitor is a *store of energy*. The energy is stored in the electric field between the plates. The energy is released when the capacitor is discharged.

▶▶ *Discharging capacitors – see pages 30–31.*

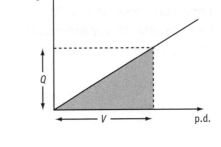

Since work done = energy transferred = charge × voltage, it follows that the energy W stored by a charged capacitor is the triangular area under the Q–V graph. Using the formula for the area of a triangle ($\frac{1}{2}$ × base × height):

energy stored = $\frac{1}{2}$ × charge × p.d.   or $W = \frac{1}{2}QV$

Substituting $Q = CV$ gives $W = \frac{1}{2}CV^2$, and substituting $V = Q/C$ gives $W = \frac{1}{2}Q^2/C = Q^2/2C$. Hence there are three forms of the equation, the first being the fundamental one:

$$W = \tfrac{1}{2}QV = \tfrac{1}{2}CV^2 = \frac{Q^2}{2C}$$

✓ *Quick check 4*

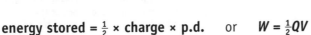

❓ Quick check questions

1 A capacitor stores 40 µC of charge when connected to a 5 V supply. What is its capacitance? How much charge will it store when connected to a 20 V supply?

2 How many 20 pF capacitors must be connected in parallel to make 100 pF?

3 What is the capacitance of three 120 µF capacitors connected in series?

4 A capacitor stores 10 mJ of energy when connected to a 100 V supply. What is its capacitance? How much energy will it store when connected to a 200 V supply?

▶ You may be able to do this a quick way, but check also that you can use the formula.

Discharging a capacitor

When a charged capacitor is disconnected from the supply used to charge it up, it can be discharged by connecting it across a resistor. The greater the resistance, the more slowly the capacitor will discharge.

Flow of charge

Current flows round from the positively charged plate through the resistor to the negatively charged plate. As the charge decreases, so does the current.

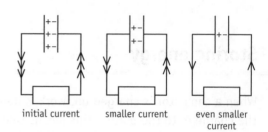

initial current   smaller current   even smaller current

Graphs

During discharge, charge Q, p.d. V and current I all follow the same pattern, an **exponential decay curve**. This curve is always getting closer to zero, without ever reaching it.

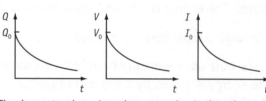

The charge stored decreases rapidly at first, then more and more slowly.

Less charge stored means less p.d. across the capacitor, so the p.d. follows the same pattern.

As the p.d. decreases, the current through the resistor must also decrease.

✓ *Quick check 1*

Discharge equations

$Q = Q_0 e^{-t/CR}$	$V = V_0 e^{-t/CR}$	$I = I_0 e^{-t/CR}$
Q_0 = initial charge (i.e. when $t = 0$)	V_0 = initial p.d.	I_0 = initial current
$Q_0 = CV_0$	$V_0 = I_0 R$	$I_0 = \dfrac{V_0}{R}$

These equations can be used to find values of Q, V and I at any time during the discharge. Make sure you know how to use the e^x function on your calculator – see the worked example.

Worked example

A 10 µF capacitor is charged up to 20 V, and then discharged through a 50 kΩ resistor. Calculate the initial current through the resistor, and the current after 2 s.

Step 1 Calculate the initial current, I_0:

$$I_0 = \frac{V_0}{R} = \frac{20 \text{ V}}{50 \times 10^3 \ \Omega} = 400 \ \mu A$$

Step 2 Using your calculator, calculate the quantity $-t/CR$, which appears in the exponential function:

$$\frac{-t}{CR} = \frac{-2}{(10 \times 10^{-6}) \times (50 \times 10^{3})} = -4$$

Step 3 Use your calculator's e^x key, then multiply by I_0:

$$I = I_0 e^{-t/CR} = 400\ \mu A \times e^{-4} = 7.3\ \mu A$$

With practice, you can merge steps 2 and 3. Always start by calculating $-t/CR$, then press e^x, then multiply by the initial value of I.

> The quantity $-t/CR$ has no units, so there is no need to include them.

✓ *Quick check 2, 3*

Time constant *CR*

The greater the value of C and the greater the value of R, the slower the discharge of a capacitor. This is because a bigger capacitor stores more charge, so it takes longer to flow away, and a bigger resistor resists the flow of charge more.

The quantity CR thus governs how quickly the capacitor discharges, and is known as the **time constant** of the circuit: symbol τ (Greek tau); units seconds (s).

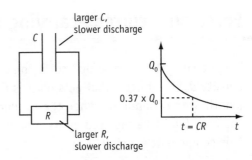

time constant τ = *CR*

The time constant is the time for the p.d. (or charge or current) to fall to $1/e$ of its initial value – that is, to about 37% of the initial value.

✓ *Quick check 4*

? *Quick check questions*

1 A capacitor is charged and discharged through a resistor. It is then charged again to the same p.d., and discharged through a resistor of twice the resistance. Which of the two graphs shown represents the second discharge? Give a reason to support your answer.

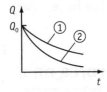

2 A 20 pF capacitor is charged up to 100 V, and then discharged through a 500 MΩ resistor. What will be the p.d. across it after 0.015 s?

3 A 12 V battery is used to charge a 1000 μF capacitor, which is then discharged through a 50 kΩ resistor. Calculate the initial current that flows, and the current after 100 s.

4 Which has the greater time constant, a circuit with a 20 μF capacitor and a 5 MΩ resistor, or a circuit with a 1000 μF capacitor and a 100 kΩ resistor?

Electromagnetic forces

An electric current has a *magnetic field* around it. If a current flows *across* another magnetic field, the two fields interact to produce a force. This is the **electromagnetic force**, *BIl*, studied in Module B. Similarly, moving charges constitute a current, and a force acts on the charges if they move through a magnetic field.

▶▶ *Force on a current-carrying conductor – see Module B, page 42.*

▶▶ *Magnetic flux density B, measured in teslas (T) – see Module B, pages 40 and 45.*

Force on a current-carrying conductor

When a current I flows at 90° to the magnetic flux (flux density B), the force F that acts on it is $F = BIl$, where l is the length of the conductor. When the conductor is at angle θ to the flux, l must be replaced by its component at 90° to B. This component is $l \sin\theta$:

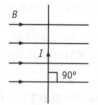 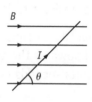

$$F = BIl \sin\theta$$

- Current flowing at 90° to flux: $\theta = 90°$, $\sin\theta = 1$, $F = BIl$.
- Current flowing parallel to flux: $\theta = 0°$, $\sin\theta = 0$, $F = 0$.

The force arises only when the current flows *across* the flux.

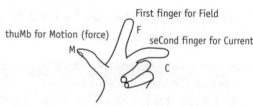

First finger for Field

thuMb for Motion (force)

seCond finger for Current

✓ *Quick check 1*

Force on a moving charge

When a positive charge Q moves with velocity v across a field of flux density B, the force F on the charge is

$$F = BQv$$

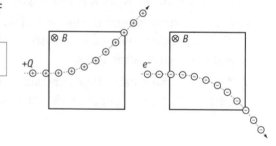

A stronger field, a greater charge and a faster charge all give a bigger force.

- Direction of the force on a *positive* charge: a current is a flow of positive charge, so use *Fleming's left-hand rule*.
- Direction of the force on a *negative* charge: the current is in the opposite direction to the movement of the charge, so point your second finger in the opposite direction to *v* when using Fleming's left-hand rule.

▶▶ *Fleming's left-hand rule – see Module B, page 42.*

✓ *Quick check 2*

Circular motion

When a charged particle moves at 90° to a magnetic field, the force on it is always at 90° to its velocity. (This is shown by Fleming's left-hand rule.) This is the condition needed for circular motion, so the charged particle will follow a circular path and we can describe the force as a *centripetal* force (see pages 15 and 16).

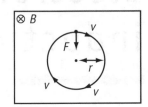

Magnetic force = mass × centripetal acceleration:

$$BQv = \frac{mv^2}{r}$$

Cancelling v from both sides and rearranging to find r gives

$$BQ = \frac{mv}{r} \quad \text{and} \quad r = \frac{mv}{BQ}$$

Looking at this equation shows that increasing the flux density will decrease the radius of the particle's orbit, i.e. the particle will go round in tighter circles.

✓ *Quick check 3–5*

? *Quick check questions*

1 A current of 5 A flows through a 2 m length of wire. The wire lies across a magnetic field of flux density 80 mT, as shown. What is the force on the wire? In which direction does the force act?

2 What force acts on an electron moving at 10^7 m s^{-1} at 90° to a magnetic field of flux density 0.1 T? (Electron charge $e = -1.6 \times 10^{-19}$ C.)

3 An electron enters a magnetic field as shown in the lower diagram. In which direction will the magnetic force on the electron act?

4 Calculate the radius of the orbit of the electron in Question 2. (Electron mass $= 9.1 \times 10^{-31}$ kg.)

5 Two electrons, moving at different speeds but in the same direction, enter a magnetic field. Which will experience the greater force? Which will move around a bigger orbit?

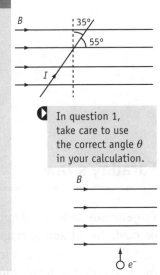

▶ In question 1, take care to use the correct angle θ in your calculation.

Electromagnetic induction

When a conductor is moved through a magnetic field, an e.m.f. may be generated across its ends. If it is part of a complete circuit, an induced current may flow. This is **electromagnetic induction**.

Flux and flux linkage

Units Flux and flux linkage are measured in **webers** (Wb). These are related to teslas (T) by

$$1 \text{ tesla} = 1 \text{ weber per square metre} \qquad 1 \text{ T} = 1 \text{ Wb m}^{-2}$$

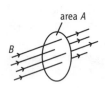

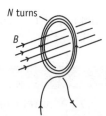

With flux density B, the flux Φ passing through area A is given by
flux $\Phi = BA$

For a coil of N turns, the *flux linkage* is N times the flux passing through it:
flux linkage = $N\Phi$ = NBA

✓ *Quick check 1, 2*

Faraday's law

To generate an induced e.m.f., a conductor must be made to cut across magnetic flux. The faster it cuts flux, the greater the induced e.m.f. For a rotating coil, the faster the flux linkage through it changes, the greater the induced e.m.f. This is **Faraday's law** of electromagnetic induction.

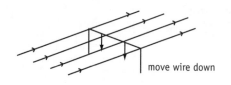

move wire down

rotate coil

In SI units,

magnitude of induced e.m.f. = rate of change of flux linkage

▶▶ *For the difference between p.d. and e.m.f., refer back to Module B, page 34.*

✓ *Quick check 3*

Lenz's law

This law determines the direction in which an induced current flows, or the polarity of an induced e.m.f. An induced current flows in a direction to oppose the change producing it. This is **Lenz's law**.

For example, if a straight conductor is moved across a magnetic field, an induced current flows in it. There is a force on this current, and this force opposes the force pushing the conductor across the field.

You can use *Fleming's right-hand rule* to determine the direction of an induced current. Thumb and fingers represent the same quantities as in the left-hand rule (Module B, page 42), but in this case using the right hand.

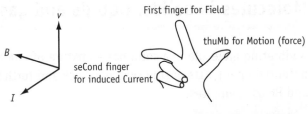

✓ *Quick check 4*

Applications of electromagnetic induction

- A coil rotating in a magnetic field (a generator)
- A magnet rotating in a coil (a bicycle dynamo)
- Transformers

❓ Quick check questions

1 2×10^{-3} Wb of magnetic flux pass through a square area 10 cm × 10 cm. What is the flux density of this field?

2 What is the flux linkage of a circular coil of radius 5 cm, consisting of 50 turns of wire, placed perpendicular to a magnetic field of flux density 200 mT?

3 A coil of wire is being rotated in a magnetic field. Which of the following will increase the e.m.f. induced across the coil: increasing the flux density of the field; increasing the rate of rotation; reversing the direction of rotation; reducing the resistance of the coil by using thicker wire?

4 The straight conductor shown is being pushed across a uniform magnetic field. Will the induced current flow towards or away from point A? Will point A acquire a positive or a negative charge?

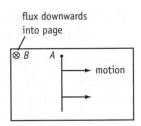

flux downwards into page

Internal energy

What happens when a substance is heated? Energy is transferred to the substance. This increases its **internal energy**. We can understand internal energy in terms of the energy of the molecules that make up the substance.

Molecules in solids, liquids and gases

A vibrating molecule in a solid has a mixture of kinetic and potential energy. Its energy switches back and forth between KE and PE as it vibrates.

A molecule moving about in a liquid or gas has **translational kinetic energy** (translational means 'moving from place to place'). It also has potential energy, because it has been separated from its former neighbours.

▶▶ *Energy of a vibrating particle – refer back to Module B, pages 48–51.*

In a solid, the molecules vibrate about fixed positions, being bonded to their neighbours.

In a liquid, the molecules can move more freely. They are more weakly bonded with their neighbours.

In a gas, molecules can move freely and rapidly. They are not bonded to one another.

✓ *Quick check 1*

Internal energy

The internal energy of an object depends on 'the state of the system' – its temperature, pressure, volume, etc. It is simply the sum of the energies (kinetic and potential) of the molecules making up the object. The energy of each molecule keeps changing in a random fashion, due to collisions with neighbours, but if we could take a snapshot of the system at an instant in time and add up the energies of all the molecules, we would find the internal energy. More formally:

The internal energy of a system is the sum of a random distribution of kinetic and potential energies associated with the molecules of the system.

Energy can be transferred to a system in many ways: by heating, by doing work, electrically, etc. The energy transferred is shared among the molecules of the system, increasing their energies. This increases the internal energy of the system.

✓ *Quick check 2, 3*

Absolute zero

As a substance is cooled, its molecules move more and more slowly. At a temperature called **absolute zero** (–273.15°C), its molecules would reach their lowest possible energy. In practice, this temperature is unattainable. At absolute zero, all substances would have a minimum internal energy.

Absolute zero is the starting point of the *Kelvin* or *thermodynamic* temperature scale (0 K, 0 kelvin): see page 41.

Changes of state

Energy must be supplied to melt a solid, or to boil a liquid. As it *changes state*, the substance remains at a steady temperature. The energy input is increasing the potential energy of the molecules, not their kinetic energy. (Bonds are broken between molecules, but they do not move any faster.)

✓ *Quick check 4*

❓ Quick check questions

1 In which state (solid, liquid or gas) do the molecules of a substance have the greatest kinetic energy? In which state the greatest potential energy?

2 A block of ice is cut exactly in half. What can you say about the internal energy of each half, compared to that of the original block?

3 Which of the following energy transfers will increase the internal energy of a block of ice? (Assume no other transfers take place.)

   • carrying it upstairs

   • heating it with an electrical heater

   • putting it in a fast-moving aircraft

4 The graph shows the results of an experiment in which a material, initially solid, is heated at a steady rate. Its temperature changes as shown. State whether the material is solid, liquid or gas at points A–E. At which of points A–E is its internal energy increasing?

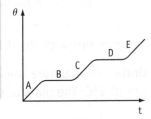

Specific heat capacity

When a substance is heated, its temperature rises (unless it changes state – see pages 40–41). Taking equal masses, some substances heat up more quickly than others – they have a low **specific heat capacity** (**s.h.c.**).

Defining s.h.c.

Energy must be supplied to raise the temperature of a substance. The amount of energy ΔQ that must be supplied depends on:

- the mass of the substance, m
- its specific heat capacity (s.h.c.), c
- the temperature rise, $\Delta\theta$

> **The specific heat capacity of a substance is the amount of energy needed to raise the temperature of 1 kg of the substance by 1°C, or by 1 K.**

▶ The word *specific* here means *per unit mass* or *per kilogram*.

✓ *Quick check 1*

Calculations

The four quantities above are related by the equation

$$\Delta Q = mc\Delta\theta$$

Rearranging this equation gives

> **specific heat capacity** $c = \dfrac{\Delta Q}{m\Delta\theta}$ **s.h.c. = energy per kg per °C**

Compare this with the definition of s.h.c. above.

Units   Temperature rise $\Delta\theta$ is measured in K (kelvin). A rise of 1 K is the same as a rise of 1°C. The units of specific heat capacity c are therefore J kg^{-1} K^{-1}.

Worked example

A 5 kg mass of water is heated electrically. A total of 210 kJ of energy is supplied. By how much will the temperature of the water rise? (Specific heat capacity of water = 4200 J kg^{-1} K^{-1}.)

Step 1   Write down what you know, and what you want to know:

$$\Delta Q = 210 \times 10^3 \text{ J}, \ m = 5 \text{ kg}, \ c = 4200 \text{ J kg}^{-1} \text{ K}^{-1}, \ \Delta\theta = ?$$

Step 2   Rearrange the formula, substitute and solve:

$$\Delta\theta = \frac{\Delta Q}{mc} = \frac{210 \times 10^3 \text{ J}}{5 \text{ kg} \times 4200 \text{ J kg}^{-1} \text{ K}^{-1}} = 10 \text{ K}$$

So the water's temperature rises by 10 K (or 10°C).

✓ *Quick check 2, 3*

Measuring s.h.c.

The electrical heater supplies energy at a steady rate to the insulated block of metal. The graph shows the rate of rise of the block's temperature. An ammeter and voltmeter are used to determine the heater's power (energy supplied per second).

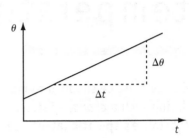

Practical points:

- Insulation helps to reduce the loss of energy from the block.

- Some energy is wasted in heating the thermometer and the heater itself.

Energy losses mean that the value of c deduced from this experiment is too high. Use the straight portion of the graph to deduce the temperature rise per second, $\Delta\theta / \Delta t$. Then:

energy supplied per second = mass × s.h.c. × temperature rise per second

✓ *Quick check 4*

❓ Quick check questions

1 The graph shows the results of an experiment in which samples of two different materials, each of mass 2 kg, were supplied with thermal energy at the same rate. Which material, A or B, has the greater specific heat capacity?

2 How much energy must be supplied to raise the temperature of a 0.5 kg block of aluminium from 20°C to 60°C? (s.h.c. of aluminium = 900 J kg^{-1} K^{-1}.)

3 A 2.5 kg block of nylon at 20°C is heated for 5 minutes using a 100 W heater. To what value will its temperature rise? (s.h.c. of nylon = 470 J kg^{-1} K^{-1}.)

4 The graph shows the results of an experiment to measure the specific heat capacity of brass. A 1 kg block was heated with a 50 W electrical heater. Use this information to estimate the specific heat capacity of brass.

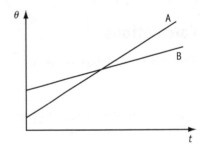

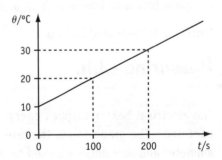

Specific latent heat; temperature scales

Energy must be supplied to melt or boil a substance. The temperature does not rise during such a *change of state*. The energy required can be calculated if the substance's **specific latent heat (s.l.h.)** is known.

Defining s.l.h.

The amount of energy ΔQ that must be supplied to melt or boil a substance depends on:

- the mass of the substance, m
- its specific latent heat (s.l.h.), L

> **The specific latent heat of a substance is the amount of energy that must be supplied to melt or boil 1 kg of the substance.**

✓ *Quick check 1*

Calculations

The three quantities above are related by the equation

$$\Delta Q = mL$$

Rearranging this equation gives

> **specific latent heat $L = \dfrac{\Delta Q}{m}$ s.l.h. = energy per kg**

Compare this with the definition of s.l.h. above.

Units L is measured in J kg^{-1}.

▶ This equation does not involve a temperature rise $\Delta\theta$, since there is no change in temperature during a change of state.

✓ *Quick check 2*

Measuring s.l.h.

The electrical heater supplies energy at a steady rate to the ice in the insulated container. The graph shows the rate of rise of the ice's temperature. An ammeter and voltmeter are used to determine the heater's power (energy supplied per second).

Practical points:

- Insulation helps to reduce the loss of energy from the ice.
- The waste of energy in heating the thermometer, the container and the heater itself is not significant, since we are concerned with the time when the temperature was steady.

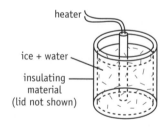

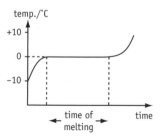

Energy losses mean that the value of L deduced from this experiment is too high. Use the horizontal portion of the graph to deduce the time during which the ice was melting. Then:

energy supplied per second × time for melting = mass × s.l.h.

Similar considerations apply to boiling.

✓ *Quick check 3*

Temperature scales

The **Celsius scale** of temperature was devised to give 100 degrees between the melting point of water (0°C) and its boiling point (100°C). Temperatures in °C below the melting point of water are therefore *negative*.

A more useful temperature scale in the study of the behaviour of matter is the **thermodynamic scale** or **Kelvin scale**. This scale has a fixed starting point: *absolute zero* (see pages 36–37). This is defined as 0 K or 0 kelvin, and temperatures, often called *absolute temperatures*, are expressed as 10 K, 200 K, etc. (note: *not* 'degrees kelvin').

◐ Remember that it is impossible to have negative temperatures on the Kelvin scale.

The unit interval on this scale, known as 1 kelvin (1 K), is also fixed – it is an SI base unit. So the thermodynamic scale does not depend on the physical properties of any particular substance. A *temperature rise* of 1 K (again, *not* 'degree kelvin') is identical with a rise of 1°C.

Absolute zero, 0 K, equals –273.15°C, though in normal work the approximation 273 is acceptable. So to convert a temperature from one scale to the other:

$$°C = K − 273$$
$$K = °C + 273$$

The usual symbols are θ for temperatures in °C and T for temperatures in K.

✓ *Quick check 4, 5*

? Quick check questions

1 Two 1 kg blocks of different metals were heated at their melting points. Block A required 100 kJ; block B required 120 kJ. Which metal had the higher specific latent heat?

2 How much energy must be supplied to vaporise 150 g of ethanol? (Specific latent heat of vaporisation of ethanol = 850 kJ kg^{-1}.)

3 A 2 kW electric kettle initially contained 800 g of water at 100°C. It boiled dry in 15.0 minutes. Estimate the specific latent heat of vaporisation of water.

4 Convert the following temperatures in °C to K: 0°C, 100°C, –50°C.

5 Which temperature is higher, 145 K or –130°C?

Gas equations

The **equation of state for an ideal gas** (also called the **ideal gas equation**) relates the pressure, volume and temperature of a gas. We can also relate these quantities to the average kinetic energy of the molecules of the gas. In these relationships, the amount of a gas is expressed in *moles*.

The mole and the Avogadro constant

The **mole** (abbreviated to **mol**) is the unit of *amount* of a substance. One mole of any substance consists of a standard number of particles. This number is N_A, the **Avogadro constant**:

$$N_A = 6.02 \times 10^{23} \text{ mol}^{-1}$$

The mass of one mole of any substance is the relative molecular mass of the substance, expressed in grams.

For example, the relative molecular mass of water is 18, so 1 mole of water has a mass of 18 g and consists of 6.02×10^{23} molecules.

✓ *Quick check 1*

The equation of state for an ideal gas

The pressure p (in pascals, Pa), volume V (m^3) and *absolute* temperature T (K) of a gas are related by

$$pV = nRT$$

where n is the number of moles of gas, and R is the **molar gas constant**:
$$R = 8.3 \text{ J K}^{-1} \text{mol}^{-1}$$

An **ideal gas** is one that obeys $pV = nRT$. In practice, most gases behave ideally only at low pressures and at temperatures well above their boiling points.

▶ Always express temperatures in kelvin (see page 41) when applying this equation.

Worked example

What volume is occupied by 1 mole of a gas at a pressure of 10^5 Pa and a temperature of 273 K?

Step 1 Write down what you know, and what you want to know:

$$n = 1 \text{ mol}, \ p = 10^5 \text{ Pa}, \ T = 273 \text{ K}, \ R = 8.3 \text{ J K}^{-1} \text{ mol}^{-1}, \ V = ?$$

Step 2 Rearrange the equation of state, substitute and solve:

$$V = \frac{nRT}{p} = \frac{1 \text{ mol} \times 8.3 \text{ J K}^{-1} \text{ mol}^{-1} \times 273 \text{ K}}{10^5 \text{ Pa}}$$

$$= 0.0227 \text{ m}^3$$

▶ If you're studying chemistry, you should recognise this as the volume of 1 mole at s.t.p.

Note that, using SI units throughout, the volume is in m^3.

✓ *Quick check 2, 3*

Molecular energy

The molecules of a gas have a range of speeds – some are moving faster than others. At higher temperatures, they move faster on average.

average KE of a molecule = $\frac{1}{2}m<c^2>$

where m is the mass of a molecule and $<c^2>$ is the *mean square speed* of the molecules (*mean* means *average*).

For N molecules, we can deduce that

$$pV = \frac{1}{3}Nm<c^2>$$

Note that this equation relates bulk properties of the whole gas (p and V on the left-hand side) to the properties of its molecules (on the right-hand side).

Comparing this equation with the equation of state $pV = nRT$ gives

$$\frac{1}{3}Nm<c^2> = nRT$$

The left-hand side is related to the average KE of the molecules; hence

average molecular KE $\propto$ absolute temperature

Hence, when we measure the absolute temperature of a gas, we are measuring the average KE of its molecules. The hotter the gas, the greater the molecular KE.

✓ *Quick check 4*

? *Quick check questions*

1 How many particles are there in 5 moles of water? In 5 moles of uranium?

2 At what temperature will 10 moles of an ideal gas occupy 0.1 m^3 at a pressure of 2 × 10^5 Pa? (Molar gas constant R = 8.3 J K^{-1} mol^{-1}.)

3 An ideal gas initially occupies a volume of 10 litres. It is compressed at a constant temperature so that its pressure increases by a factor of 2.5. Calculate its new volume.

4 A gas is heated from 300 K to 600 K at a constant pressure. What happens to the average kinetic energy of its molecules during this process? The gas is then compressed to half its original volume, still at 600 K. What happens to the average KE of the molecules during this second process?

The structure of matter

Solid materials are made of **atoms**, closely packed together. In some materials, the atoms are arranged in a regular pattern – they have a **crystalline** structure. The underlying structure of materials can be revealed by a variety of **diffraction** techniques.

X-ray diffraction

A beam of monochromatic X-rays (all one wavelength) is directed at a crystal. The X-rays are diffracted by the regularly spaced planes of atoms in the crystal. A diffraction pattern of spots is detected using photographic film or an electronic detector. The arrangement and spacing of the spots tell us about:

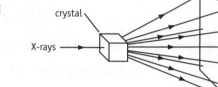

- the arrangement of the atomic planes (**crystal structure**);
- the separation of the atomic planes.

Rotating the crystal to a different orientation gives a different pattern, because the spacing of planes is different.

A *polycrystalline* material (many tiny crystals) gives a pattern of rings, not dots. An *amorphous* material (disordered structure) gives a pattern of blurred rings.

▶▶ *Diffraction of waves – refer back to Module C, pages 68–69.*

▶▶ *X-rays as electromagnetic waves – see Module B, pages 46–47.*

✓ Quick check 1

Electron and neutron diffraction

Instead of X-rays, beams of *electrons* or *neutrons* can be used. *Wave–particle duality* tells us that moving particles have a wavelength associated with them. If this is similar to the spacing of the atomic planes, the beam will be diffracted. These techniques give similar information to X-ray diffraction concerning the arrangement and spacing of the atomic planes.

(Electrons interact with the atomic planes because they are electrically charged. Neutrons are uncharged, but they are magnetic, so they interact with any magnetic atoms in the material.)

▶▶ *Wave–particle duality and electron diffraction – refer back to Module B, pages 52–53.*

✓ Quick check 2

Alpha-particle scattering

An atom is neutral – it contains equal amounts of positive and negative charge. The positive charge is concentrated in the tiny *nucleus* at the centre. The negative charge (of the electrons) is spread out around the nucleus. Evidence for this comes from Rutherford's **alpha-particle scattering experiment**. (Alpha-particles are positively charged helium nuclei, consisting of 2 protons and 2 neutrons.)

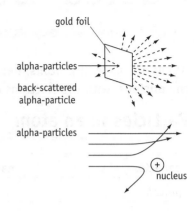

evidence	deduction
Most alpha-particles pass straight through gold foil.	The atoms of the gold foil are mostly 'empty space'.
A few (1 in about 10^4) are deflected back towards the observer.	Positive charge is concentrated in a tiny volume – the nucleus.

High-energy electron scattering experiments use beams of very fast-moving electrons. These have extremely short wavelengths, similar to the size of the nucleus. The resulting diffraction patterns are interpreted to give the diameter of the nucleus.

✓ *Quick check 3*

Relative sizes

- Diameter of nucleus ~10^{-15} m (1 fm)
- Diameter of atom ~10^{-10} m (0.1 nm)

A **molecule** may be similar in size to a single atom, or it may be much larger, depending on how many atoms it is made of. A protein molecule may have a diameter of 10^{-7} m.

✓ *Quick check 4, 5*

Quick check questions

1. In X-ray diffraction, why must the wavelength of the X-rays be similar to the spacing of the atomic planes?
2. If the energy of electrons in a beam is increased, how does their momentum change, and their associated wavelength?
3. In the alpha-particle scattering experiment, Rutherford used gold foils of different thicknesses. Explain how changing the thickness of the foil would change the number of alpha-particles back-scattered.
4. By how many orders of magnitude (factors of 10) does the diameter of an atom exceed that of its nucleus?
5. The smallest object resolvable using an optical microscope has a diameter of the order of 1 µm. Roughly how many atomic diameters is this?

Nuclear structure

Atoms are made of **protons**, **neutrons** and **electrons**. Protons and neutrons make up the **nucleus**, with the electrons orbiting around it.

Particles in an atom

particle	mass	charge
proton	1	+1
neutron	~1	0
electron	~1/1840	−1

Protons and neutrons are **nucleons** (particles found in the nucleus).

Masses are given relative to the proton. The mass of a neutron is very slightly more than that of a proton.

Charges are in units of $e = 1.6 \times 10^{-19}$ C.

Representing a nucleus

- Z = **proton number** (or **atomic number**) = number of protons in nucleus
- A = **nucleon number** (or **mass number**) = number of nucleons in nucleus
- N = **neutron number** = number of neutrons in nucleus

Since protons and neutrons are nucleons,

$$A = Z + N$$

In a neutral atom, number of electrons = number of protons = Z.

An individual nucleus can be represented by $^A_Z X$, e.g. $^{16}_{8}O$ and $^{238}_{92}U$.

- Upper number A = nucleon number
- Lower number Z = proton number

Each combination of A and Z represents a different nuclear species or **nuclide**.

Alpha- and beta-particles can also be represented in this way:

- **alpha-particle** (a helium nucleus – 2 protons and 2 neutrons): ^4_2He
- **beta-particle** (an electron – zero mass, negative charge): $^0_{-1}e$

✓ *Quick check 1, 2*

Isotopes

Most elements come in a variety of forms or **isotopes**. Each isotope has the same number of protons in the nucleus, but different numbers of neutrons, so their masses are different.

$^1_1H \quad ^2_1H$

two isotopes of hydrogen

$^{54}_{26}Fe \quad ^{56}_{26}Fe$

two isotopes of iron

✓ *Quick check 3*

Representing nuclear processes

Alpha decay A nucleus emits an alpha-particle, i.e. a helium nucleus (2 protons and 2 neutrons). For example:

$$^{226}_{88}\text{Ra} \rightarrow {}^{222}_{86}\text{Rn} + {}^{4}_{2}\text{He}$$

Beta decay A nucleus emits a beta-particle, i.e. an electron (zero mass, negative charge). For example:

$$^{14}_{6}\text{C} \rightarrow {}^{14}_{7}\text{N} + {}^{0}_{-1}\text{e}$$

Gamma decay A gamma-photon (electromagnetic radiation, represented by γ) may also be emitted during alpha or beta decay. For example:

$$^{241}_{95}\text{Am} \rightarrow {}^{237}_{93}\text{Np} + {}^{4}_{2}\text{He} + \gamma$$

✓ *Quick check 4*

Charge and mass conservation

From the above equations, you can see that:

● charge is conserved (same total of Z's on left and right);

● nucleon (mass) number is conserved (the A's also balance).

▶▶ *More about mass conservation on pages 48–49.*

? *Quick check questions*

1 Represent in symbolic form a nucleus of a silicon (Si) atom that consists of 14 protons and 14 neutrons.

2 How many protons, neutrons and electrons are there in a neutral oxygen atom whose nucleus is represented by $^{16}_{8}\text{O}$?

3 The table shows the composition of four nuclei. Which nuclei are isotopes of the same element?

nucleus	number of protons	number of neutrons
A	24	26
B	23	26
C	23	27
D	24	27

4 Write equations to represent the following nuclear decays:

● A polonium nucleus $^{210}_{84}\text{Po}$ emits an alpha-particle and a gamma-photon to become an isotope of lead Pb.

● A potassium nucleus $^{42}_{19}\text{K}$ decays by beta emission to become an isotope of calcium Ca. A gamma-photon is also emitted.

Mass–energy conservation

Energy is released during radioactive decay. The particles fly apart (they have kinetic energy), and a gamma-photon may be released. Where does this energy come from?

Mass and energy units

The masses of subatomic particles are very small; it is convenient to express them in **atomic mass units** (**u**), rather than kg:

$$1 \text{ u} = 1.66 \times 10^{-27} \text{ kg}$$

- Rest mass of proton = 1.67×10^{-27} kg = 1.007 u
- Rest mass of electron = 9.11×10^{-31} kg = 0.000 549 u

Similarly, energy may be given in **electronvolts** rather than joules:

$$1 \text{ eV} = 1.6 \times 10^{-19} \text{ J}$$

▶▶ *Electronvolts – refer back to Module B, page 48.*

✓ *Quick check 1*

Disappearing mass

In radioactive decay, a nucleus emits one or more particles.

mass of particles before decay > mass of particles after decay

The decrease in mass Δm is accounted for by the appearance of an amount of energy ΔE. Δm and ΔE are related by **Einstein's equation**:

$$\Delta E = \Delta m \times c^2$$

where c is the speed of light in free space, 3×10^8 m s^{-1}.

Most people remember this equation as $E = mc^2$.

Worked example

A neutron decays to become a proton and an electron. How much energy is released? (Values of rest mass are shown in the table. Inspection shows that the proton and electron together have less mass than the neutron.)

particle	rest mass / kg
neutron, n	$m_n = 1.674\ 928 \times 10^{-27}$
proton, p	$m_p = 1.672\ 623 \times 10^{-27}$
electron, e	$m_e = 0.000\ 911 \times 10^{-27}$

Step 1 Write down an equation for the reaction:

$$n \rightarrow p + e$$

Step 2 Calculate the loss in mass Δm:

$$\Delta m = m_n - (m_p + m_e) = 1.394 \times 10^{-30} \text{ kg}$$

Note that it is necessary to work to 7 significant figures, because the differences between the quantities are very small. Use your calculator!

Step 3 Calculate the energy ΔE released:

$$\Delta E = \Delta m \times c^2 = 1.394 \times 10^{-30} \text{ kg} \times (3 \times 10^8 \text{ m s}^{-1})^2 = 1.25 \times 10^{-13} \text{ J}$$

✓ *Quick check 2*

Applying $\Delta E = \Delta m \times c^2$

The idea of energy conservation has been replaced by the idea of **mass–energy conservation**. Einstein's equation applies in all energy changes; however, it is most significant in nuclear changes.

The **rest mass** of a particle is its mass when at rest. A moving particle has kinetic energy, and if we could measure its mass, we would find it was greater than its rest mass.

✓ *Quick check 3*

Binding energy, fission and fusion

The nucleons of a nucleus are bound together. Energy is needed to separate them. The **binding energy** of a nucleus is the energy needed to separate a nucleus into its individual nucleons. (It may help to think of this as the *unbinding energy* of the nucleus.) The graph shows the binding energy per nucleon of all nuclei. Iron (Fe) is at the highest point; its nucleons are most tightly bound together.

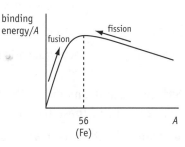

In **nuclear fusion**, small nuclei bind together to form a bigger nucleus. Energy is released; the nucleons are more tightly bound together.

In **nuclear fission**, a large nucleus splits to form two or more smaller nuclei. Again, energy is released; the nucleons are more tightly bound together in the resulting nuclei.

✓ *Quick check 4*

❓ Quick check questions

1   The rest mass of a neutron is $1.674\,928 \times 10^{-27}$ kg. What is this in atomic mass units?

2   A fast-moving electron may be captured by a proton to form a neutron. However, a stationary electron cannot be captured. Explain why not, using ideas about mass–energy conservation. (You may wish to refer to the table of mass values on the opposite page.)

3   The Sun radiates energy into space at the rate of 4×10^{26} W. By how much does its mass decrease each second? ($c = 3 \times 10^8$ m s^{-1}.)

4   In nuclear fission, a 'mother' nucleus splits to form two 'daughter' nuclei; some neutrons are also released. What can you say about the mass of these products, compared to that of the mother nucleus?

Radioactive decay

It is impossible to predict when an individual nucleus will decay; it occurs spontaneously. As a consequence, the decay of a sample is *random*. If you monitor the decay of a long-lived isotope, you will observe fluctuations in the count rate. Consequently, readings must be averaged.

Ionising radiation

As alpha, beta and gamma radiations pass through matter, they lose energy. Their interactions with matter cause ionisation – electrons are knocked from neutral atoms. Alpha radiation is the most strongly interacting, so its range is least.

radiation	nature	absorbed by	penetrating power
alpha, α	helium nucleus (2 protons + 2 neutrons)	thin paper	least
beta, β	electron	a few mm of aluminium	
gamma, γ	electromagnetic radiation	a few cm of lead	most

Safety precautions in handling, storage and disposal of radioactive materials must take account of these hazards.

● Avoid direct contact with sources.

● Handle solid sources using tongs.

● Keep at a safe distance when working with sources.

● Store sources in lead-lined containers.

● Dispose of sources away from human activity, and in forms that cannot leak.

✓ Quick check 1

Activity of a sample

The **activity** A of a sample is the average number of nuclei in the sample that decay per second (the average rate of decay).

Units   Activity is expressed in **becquerels (Bq)**:

$$1 \text{ Bq} = 1 \text{ s}^{-1} \text{ (1 decay per second)}$$

We have to consider the *average* (or *mean*) number of decays per second, because the rate fluctuates randomly. In practice, it is difficult to detect all the decays that occur, so we may use the measured **count rate**, instead of the activity. The **corrected count rate** takes account of the background count rate.

✓ Quick check 2

Decay constant

Some radioactive materials decay very quickly; others decay very slowly. The difference lies in the **decay constant** λ. The decay constant for a particular isotope is the probability that an individual nucleus will decay in unit time.

Units s^{-1} (or day^{-1}, or $year^{-1}$, etc.)

For example, suppose an isotope has a decay constant $\lambda = 0.1$ $year^{-1}$. If we could observe a single nucleus of this isotope for one year, there is a probability of 0.1 (i.e. 1 in 10) that it will decay in this time. If we could observe 100 of these nuclei for a year, we would expect roughly 10 to decay.

Hence, the activity A of a sample depends on two things:

- the decay constant λ of the isotope;
- the number of undecayed nuclei N it contains.

activity = decay constant × number of undecayed nuclei $A = \lambda N$

Worked example

A sample of radioactive carbon contains 2×10^{13} undecayed carbon nuclei. The decay constant for this isotope is 4×10^{-12} s^{-1}. What is the activity of the sample?

Substituting in the equation gives

$$A = \lambda N = 4 \times 10^{-12} \text{ } s^{-1} \times 2 \times 10^{13} = 80 \text{ } s^{-1}$$

So the activity is 80 s^{-1}, or 80 Bq.

✓ *Quick check 3*

? *Quick check questions*

1 Of the three types of ionising radiation considered here (α, β and γ), which interacts least strongly with matter as it passes through it?

2 A Geiger counter placed next to a sample of a radioactive material detects an average of 1.5 counts per second. Give two reasons why it would be incorrect to conclude that the sample's activity is 1.5 Bq.

3 A sample of a radioactive isotope contains 5×10^8 undecayed nuclei. Its activity is 600 Bq. What is the decay constant for this isotope?

Radioactive decay equations

The decay of a radioactive substance is like the discharge of a capacitor: both follow an exponential pattern. The rate of decay depends on the **half-life** of the radioactive substance. The half-life is related to the decay constant.

The pattern of decay

As a sample of a radioactive substance decays, several quantities follow the same pattern:

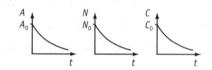

- A, the activity of the sample
- N, the number of undecayed nuclei
- C, the corrected count rate.

Each of these quantities starts at a certain initial value (e.g. A_0); its value falls rapidly at first, then more and more slowly. This is an **exponential decay**.

In practice, because radioactive decay is random and spontaneous, an experimental curve will have points scattered about the smooth theoretical curve. Also, a true exponential curve never reaches zero. However, in the case of radioactivity, the last undecayed nucleus may eventually decay, and the curve will reach zero.

✓ *Quick check 1*

The decay equation

Each of the quantities above follows an equation of the same form:

$$x = x_0 e^{-\lambda t}$$

For example, activity A varies as $A = A_0 e^{-\lambda t}$.

✓ *Quick check 2*

Worked example

A sample of a radioactive nuclide initially consists of 3×10^6 undecayed nuclei. How many will remain undecayed after 1 hour? The decay constant λ for this nuclide is $10^{-3}\ s^{-1}$.

Step 1 Write down what you know, and what you want to know:

$$N_0 = 3 \times 10^6,\ \lambda = 10^{-3}\ s^{-1},\ t = 1\ h = 3600\ s,\ N = ?$$

▶ Note that t must have the same units as $1/\lambda$.

Step 2 Write down the appropriate form of the decay equation:

$$N = N_0 e^{-\lambda t}$$

Step 3 Substitute and solve, calculating the value of $-\lambda t$ first:

$$N = 3 \times 10^6 \exp(-10^{-3}\ s^{-1} \times 3600\ s) = 3 \times 10^6 \exp(-3.6) = 82\ 000$$

You should be able to complete this calculation without writing down the intermediate step. Note that the answer is given to 2 significant figures; because of the randomness of radioactive decay, we cannot say that 81971 nuclei will remain after 1 hour.

✓ *Quick check 3*

Half-life and decay constant

The **half-life** $t_{1/2}$ of a radioactive nuclide is the mean (average) time for the number of nuclei of that nuclide to decay to half of its original value.

We have to say *mean time* since every measurement will give a slightly different value, because of random fluctuations.

The smaller the decay constant λ, the longer the half-life $t_{1/2}$. These two quantities are related by

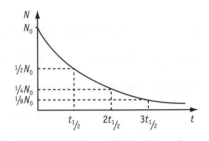

$$\lambda t_{1/2} = 0.693$$

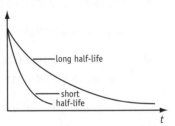

✓ *Quick check 4, 5*

? *Quick check questions*

1 Use the equation $A = \lambda N$ to explain why the activity A of a radioactive nuclide follows the same pattern of decay as the number of undecayed nuclei N.

2 Write down an exponential decay equation to represent how the corrected count rate C decreases with time t.

3 A particular radioactive nuclide has decay constant 0.03 year^{-1}. A sample has initial activity 40 Bq. What will its activity be after 5 years?

4 Radioactive carbon-14 has a half-life of 5570 years. What is the decay constant for this isotope?

5 Nuclide X has decay constant 6×10^{-3} s^{-1}. What is its half-life? What fraction of a sample of this nuclide will remain after 200 s?

Module D: end-of-module questions

Acceleration due to gravity $g = 9.8$ m s^{-2}

Molar gas constant $R = 8.31$ J K^{-1} mol^{-1}

Avogadro constant $N_A = 6.02 \times 10^{23}$ mol^{-1}

1 a Define (linear) momentum and state whether it is a vector or scalar quantity.

b Calculate the momentum of a bus of mass 6000 kg moving at 20 m s^{-1}.

c State the principle of conservation of momentum.

d Explain how the principle of conservation of momentum applies in the following situations:

(i) a child, initially stationary, jumps up in the air;

(ii) the child lands on the ground (without bouncing).

2 a What is meant by an *elastic* collision?

b A marble of mass 5 g moving at 1 m s^{-1} collides with an identical, stationary marble. The first marble stops dead and the second moves off at 1 m s^{-1}. Show that this collision is elastic.

3 The diagram shows a ball rolling down a hill. Three forces act on it as shown:

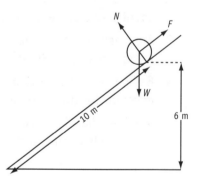

• the ball's weight $W = 1$ N

• the frictional force $F = 0.5$ N

• the normal reaction of the slope, $N = 0.8$ N

Calculate the work done by each force as the ball rolls down the slope.

4 A fruit of mass 100 g falls from a branch of a tree, 12.0 m above the ground. It hits the ground with a speed of 14.8 m s^{-1}. Calculate:

a the change in the fruit's gravitational potential energy during the fall;

b the change in its kinetic energy.

c With reference to the principle of conservation of energy, state the energy transfers that occur as the fruit falls.

5 a An electron follows a circular orbit of radius r around the nucleus of an atom. Draw a diagram to represent this. Include arrows to show:

• the electron's velocity v;

• the centripetal force F acting on the electron.

b Write down an expression for F in terms of v, r and the electron's mass m.

6 a Jupiter has mass M; one of its moons has mass m and orbits along a circular path of radius r. Write down an expression for the gravitational force F that Jupiter exerts on its moon. What can you say about the gravitational force exerted by this moon on Jupiter?

b The moon orbits Jupiter with speed v. Use the data below to calculate v:

- Mass of Jupiter $M = 1.9 \times 10^{27}$ kg

- Radius of moon's orbit $r = 7.0 \times 10^5$ km

- Universal gravitational constant $G = 6.67 \times 10^{-11}$ N m^2 kg^{-2}

c Calculate the duration of one complete orbit of Jupiter.

7 A small mass hangs from the end of a light spring. It is displaced slightly downwards and released. It oscillates up and down with simple harmonic motion. Its oscillations are represented by the graph.

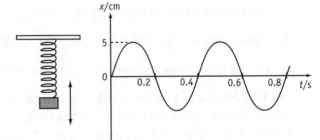

a State what is meant by *simple harmonic motion*.

b What are the amplitude, period and frequency of the oscillations?

c Write down an equation to represent the mass's displacement x as a function of time t. Use your equation to deduce the mass's displacement when $t = 20$ s.

d Calculate the maximum value of the mass's acceleration. At what point in its oscillation does it have this value?

8 The diagram shows the electric field between two positively charged metal spheres.

a Explain how a diagram of this sort represents the direction of the electric field at a point, and the strength of the field.

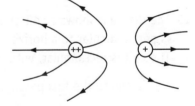

b Add arrows to the diagram to show the electric force each sphere exerts on the other. What can you say about the magnitudes and directions of these two forces?

9 Two parallel plates, 10 cm apart, are connected to a 2 kV power supply. A dust particle with positive charge 1×10^{-15} C is midway between them.

a Draw a diagram to represent the electric field between the plates.

b Calculate the electric field strength in the gap between the plates.

c Calculate the electric force on the dust particle.

d If the particle moves 3 cm closer to the positively charged plate, what can you say about the electric force on it?

10 A 10 μF capacitor is charged to 400 V.

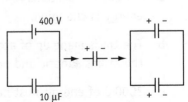

a Calculate the charge stored by the capacitor.

b Calculate the energy it stores.

The capacitor is disconnected from the source of p.d., and connected to an identical, uncharged capacitor. The charge stored by the capacitor is now shared between the two.

c Are the capacitors connected in series or in parallel? Calculate their combined capacitance.

d Calculate the p.d. across each of the capacitors.

e Calculate the energy stored by each capacitor. What fraction of the energy stored by the single capacitor is this?

11 A 2000 μF capacitor is charged up until the p.d. between its plates is 100 V. It is then allowed to discharge through a 500 kΩ resistor.

a Calculate the initial current that flows through the resistor.

b Sketch a graph to show how the current through the resistor changes.

c Calculate the time constant for the circuit.

d Write down an equation to represent how I depends on time t.

e Calculate the current flowing through the resistor after 500 s.

12 The flux density of a magnetic field can be found by measuring the force on a current-carrying conductor placed in the field. In such a measurement, a 20 cm length of conductor carrying a current of 1.5 A is placed in a field; the greatest force acting on the conductor is found to be 0.06 N.

a Draw a diagram to show the relative orientations of the conductor and the magnetic field when the force has its maximum value.

b Calculate the flux density of the field.

c The conductor is then turned through an angle of 60°. Calculate the force that now acts on it.

13 The diagram shows a coil of 100 turns of wire, placed so that its plane is perpendicular to a horizontal magnetic field of flux density 0.05 T. The coil is rectangular, with sides of lengths 6 cm and 10 cm.

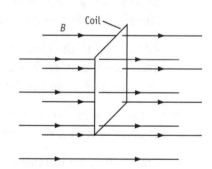

a Calculate the flux passing through the coil, and its flux linkage.

b The coil is slowly moved downwards through the field, so that its plane remains perpendicular to the flux. Explain why no e.m.f. is induced between its ends.

c How could the coil be moved to induce an e.m.f.?

14 A sample of tin is heated from 10°C below its melting point to 10°C above its melting point. The graph shows how its temperature changes with time.

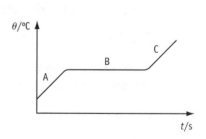

a For each of sections A–C of the graph, state how the tin's internal energy is changing.

b The tin is made up of atoms. For each of sections A–C, state how the atoms' kinetic and potential energies are changing.

c 3000 J of energy must be supplied to melt a 50 g sample of tin (at its melting point). Calculate the specific latent heat of melting of tin.

15 a One mole of oxygen consists of 6.02×10^{23} molecules. Of what quantity is the mole the unit?

b Write down an equation linking the pressure p and volume V of 1 mole of an ideal gas to its absolute temperature T.

c Use your equation to calculate the pressure of 1 mole of oxygen at 100°C if it occupies a volume of 0.025 m^3.

16 a The equation $p = Nm<c^2>/3V$ relates the pressure p and volume V of a gas to its molecular properties. Explain the meanings of each of the symbols N, m and $<c^2>$.

b A sample of 1 mole of a gas occupies a volume of 0.05 m^3 when its pressure is 5×10^4 Pa. Calculate the average kinetic energy of its molecules.

17 Here are three techniques used to investigate the structure of matter:

- X-ray diffraction

- alpha-particle scattering

- high-energy electron scattering.

Name one or more of these techniques that give evidence of:

a the *existence* of the atomic nucleus;

b the *size* of the atomic nucleus;

c the crystal structure of matter.

18 The graph shows how the binding energy per nucleon depends on atomic number for different nuclei. Use the graph to explain why the process of nuclear fission results in a release of energy.

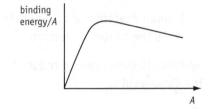

binding energy/A

A

19 a A radioactive nuclide of potassium is represented by the symbol $^{42}_{19}K$. In the nucleus of such an atom, how many protons are there, and how many neutrons?

b This nuclide decays by beta decay to an isotope of calcium (Ca). Write an equation to represent this decay.

c The half-life of this isotope of potassium is 12.5 h. Use the relationship $\lambda t_{1/2} = 0.693$ to calculate its decay constant.

20 A sample of a radioactive substance has an initial activity of 50 Bq. Its decay constant is 3×10^{-3} s^{-1}.

a Write an equation of the form $x = x_0 e^{-\lambda t}$ to represent how the activity of the sample will change with time.

b What will the sample's activity be after 500 s?

c A Geiger counter is held near the sample. The count rate detected by the counter is less than the sample's activity. State *two* factors that contribute to this.

Appendix 1: Accuracy and errors

Physicists try to make their observations as accurate as possible. Errors in measurements arise in a number of ways and, as an experimentalist, you should try to minimise errors.

Systematic errors

These can arise in a number of ways:

- **Zero error:** e.g. an ammeter does not read zero when no current is flowing through it. If it reads +0.05 A, all of its readings will be too high. Either correct the meter to read zero, or adjust all readings to take account of the error.

- **Incorrect calibration** of an instrument: e.g. an ammeter that reads zero when no current flows, but all other readings are consistently too low or too high. It may read 9.9 A when 10.0 A is flowing. Again, either correct the meter, or adjust all readings.

- **Incorrect use** of an instrument: e.g. screwing a micrometer too tightly, or viewing a meniscus from an angle. Learn the correct technique for using instruments and apparatus.

- **Human reaction:** e.g. when starting and stopping a stopclock. You may always press the button a fraction of a second after the event.

Systematic errors can be reduced or even eliminated. This increases the accuracy of the final result.

Random errors

These often arise as a result of judgements made by the experimenter:

- **Reading from a scale.** You may have to judge where a meter needle is on a scale – what is the nearest scale mark? What fraction of a division is nearest to the needle?

- **Timing a moving object.** When did it start to move? When did it pass the finishing line? You have to judge.

The conditions under which the measurement is made can vary:

- **Equipment** can vary. One trolley may have more friction than another. Two apparently identical resistors may have slightly different values.

- **Samples of materials** may be different. Two lengths of wire from the same reel may have slightly different compositions.

- **Conditions** can vary. Room temperature may change and affect your results.

Some measurements are intrinsically random:

- **Radioactive decay.** If you measure the background radiation in the laboratory for 30 s, you are likely to find slightly different values each time.

Random errors can be reduced, but it is usually impossible to eliminate them entirely. Reducing random errors increases the precision of the final result.

Reducing random errors

Here are some ways to reduce random errors.

- **Make multiple measurements**, and find the mean (average). Roughly speaking, taking four measurements reduces the error by half; 100 measurements will divide the error by 10.
- **Plot a graph**, and draw a smooth curve or a straight line through the points.
- **Choose a suitable instrument** to reduce errors of judgement, e.g. using light gates and an electronic timer instead of timing with a stopwatch. You need to think critically about the instrument: does it introduce other sources of error?

Expressing errors

Here are two ways in which the error or uncertainty in a final result can be expressed.

- **Use significant figures:** a calculation may give $R = 127\ \Omega$. If the errors are small, you may wish to quote this as $130\ \Omega$; if the errors are large, as $100\ \Omega$.
- **Use ± errors:** by considering the errors in individual measurements, you may be able to show the degree of uncertainty in the above result. Small error: $R = (127 \pm 2)\ \Omega$; larger error: $R = (130 \pm 10)\ \Omega$.

Summary

- Think critically about the equipment and methods you use.
- Reduce random errors to increase the precision of your results.
- Reduce systematic errors to increase the accuracy of your results.
- Indicate the extent of error or uncertainty in individual results, and in the final result.

Appendix 2: Data and formulae for question papers

In end-of-module question papers, you will be supplied with a long list of data and formulae. **Part 1** below shows the complete list of data and formulae included in all papers.

You are expected to recall many formulae. **Part 2** below shows the complete list, grouped together according to the module in which they are introduced.

You will not be expected to use formulae that have been introduced in an optional topic you have not studied.

Part 1: Data and formulae supplied in question papers

Data

acceleration of free fall, $g = 9.81$ m s^{-2}

gravitational constant, $G = 6.67 \times 10^{-11}$ N m^2 kg^{-2}

speed of light in free space, $c = 3.00 \times 10^8$ m s^{-1}

Planck constant, $h = 6.63 \times 10^{-34}$ J s

elementary charge, $e = 1.60 \times 10^{-19}$ C

rest mass of electron, $m_e = 9.11 \times 10^{-31}$ kg

rest mass of proton, $m_p = 1.67 \times 10^{-27}$ kg

unified atomic mass constant, $u = 1.66 \times 10^{-27}$ kg

permeability of free space, $\mu_0 = 4\pi \times 10^{-7}$ H m^{-1}

permittivity of free space, $\varepsilon_0 = 8.85 \times 10^{-12}$ F m^{-1}

molar gas constant, $R = 8.31$ J K^{-1} mol^{-1}

Avogadro constant, $N_A = 6.02 \times 10^{23}$ mol^{-1}

Formulae

uniformly accelerated motion, $s = ut + \frac{1}{2}at^2$

$v^2 = u^2 + 2as$

refractive index, $n = \dfrac{1}{\sin C}$

capacitors in series, $\dfrac{1}{C} = \dfrac{1}{C_1} + \dfrac{1}{C_2} + ...$

capacitors in parallel, $C = C_1 + C_2 + ...$

capacitor discharge, $x = x_0 e^{-t/CR}$

pressure of an ideal gas, $p = \dfrac{Nm <c^2>}{3V}$

radioactive decay, $x = x_0 e^{-\lambda t}$

half-life, $t_{1/2} = \dfrac{0.693}{\lambda}$

Option E1
critical density of matter in the Universe, $\rho_0 = \dfrac{3H_0^2}{8\pi G}$

relativity factor, $\gamma = \sqrt{(1 - v^2/c^2)}$

Option E2
sound intensity level, $I.L. = 10 \log (I/I_0)$

Option E3
current, $I = nAve$

Option E4
nuclear radius, $r = r_0 A^{1/3}$

It is a good idea to become familiar with these lists of data and formulae, so that you know what is provided in the question papers.

Part 2: Formulae not supplied in question papers

Module A: Forces and motion

speed, $v = \dfrac{s}{t}$

acceleration, $a = \dfrac{v - u}{t}$

force, $F = ma$

weight, $W = mg$

density, $\rho = \dfrac{m}{V}$

moment of a force, $T = Fx$

torque of a couple, $T = Fx$

pressure, $p = \dfrac{F}{A}$

work done, $W = Fd$

kinetic energy, $E_k = \frac{1}{2}mv^2$

gravitational potential energy, $\Delta E_p = mg\Delta h$

power, $P = \dfrac{W}{t} = Fv$

stress, $\sigma = \dfrac{F}{A}$

strain, $\varepsilon = \dfrac{\Delta l}{l}$

Young modulus, $E = \dfrac{\text{stress}}{\text{strain}} = \dfrac{\sigma}{\varepsilon}$

Module B: Electrons and photons

electric current, $I = \dfrac{\Delta Q}{\Delta t}$

potential difference, $V = \dfrac{W}{Q} = \dfrac{P}{I}$

electrical resistance, $R = \dfrac{V}{I}$

resistivity, $\rho = \dfrac{RA}{l}$

power, $P = VI = I^2 R = \dfrac{V^2}{R}$

electrical energy, $W = VIt$

resistors in series, $R = R_1 + R_2 + \ldots$

resistors in parallel, $\dfrac{1}{R} = \dfrac{1}{R_1} + \dfrac{1}{R_2} + \ldots$

force on a current-carrying conductor, $F = BIl \sin \theta$

photon energy, $E = hf$

photo-electric effect, $hf = \phi + E_{k\,max} = \phi + \tfrac{1}{2}mv^2_{max}$

de Broglie equation, $\lambda = \dfrac{h}{p} = \dfrac{h}{mv}$

Module C: Wave properties

refractive index, $n = \dfrac{c_i}{c_r} = \dfrac{n_r}{n_i}$

$n = \dfrac{\sin i}{\sin r}$

wave speed, $v = f \lambda$

double-slit interference, $\lambda = \dfrac{ax}{D}$

Module D: Forces, fields and energy

momentum, $p = mv$

force, $F = \dfrac{\Delta p}{\Delta t}$

centripetal acceleration, $a = \dfrac{v^2}{r}$

simple harmonic motion, $a = -(2\pi f)^2 x$

$x = A \sin 2\pi ft$

$x = A \cos 2\pi ft$

Newton's law of gravitation, $F = \dfrac{Gm_1 m_2}{r^2}$

gravitational field strength, $g = \dfrac{F}{m}$

$g = \dfrac{GM}{r^2}$

Coulomb's law, $F = \dfrac{Q_1 Q_2}{4\pi \varepsilon_0 r^2}$

electric field strength, $E = \dfrac{F}{Q}$

$E = \dfrac{Q}{4\pi\varepsilon_0 r^2}$

$E = \dfrac{V}{d}$

capacitance, $C = \dfrac{Q}{V}$

energy of charged capacitor, $W = \frac{1}{2}QV = \frac{1}{2}CV^2 = \dfrac{\frac{1}{2}Q^2}{C}$

time constant of CR circuit, $\tau = CR$

force on moving charged particle, $F = Bqv \sin \theta$

magnetic flux, $\Phi = BA$

induced e.m.f., $E = \dfrac{N\Delta\Phi}{\Delta t}$

ideal gas equation, $pV = nRT$

thermal energy change (s.h.c.), $\Delta Q = mc\Delta\theta$

thermal energy change (s.l.h.), $\Delta Q = mL$

mass–energy, $\Delta E = \Delta m \times c^2$

radioactivity, $A = \lambda N$

Module E1: Cosmology

apparent magnitude, $m = -2.5 \log I + \text{constant}$

apparent/absolute magnitude, $m - M = 5 \log (r/10)$

Hubble's law, $v = H_0 d$

age of Universe, $t \sim \dfrac{1}{H_0}$

Doppler formula, $\dfrac{\Delta\lambda}{\lambda} = \dfrac{v}{c}$

Module E2: Health Physics

lens formula, $\text{power} = \dfrac{1}{f} = \dfrac{1}{u} + \dfrac{1}{v}$

X-ray attenuation, $I = I_0 e^{-\mu x}$

Module E3: Materials

Hall voltage, $V_H = Bvd$

Module E5: Telecommunications

inverting amplifier gain, $G = -\dfrac{R_F}{R_{IN}}$

power ratio, $\text{number of decibels (dB)} = 10 \log (P_1/P_2)$

Appendix 3: Useful definitions

Here are brief statement of the definitions you need to know. They have been grouped together in clusters of related terms, to help you learn them.

Module D: Forces, fields and energy

Block D1: Newton's laws

linear momentum
The product of an object's mass and linear velocity.

conservation of momentum
When objects interact, the (vector) total of their momentum remains constant, provided no external force acts (i.e. in a closed system).

force
A force is equal to the rate of change of momentum of the object on which it acts.

Block D2: Oscillations and circular motion

simple harmonic motion
Motion when the force on an object is proportional to the object's displacement from a fixed point, and is always directed towards a fixed point.

critical damping
The degree of damping required for the displacement of a system to reach a constant value in the minimum time without oscillation.

Block D3: Force fields

Newton's law of gravitation
The gravitational force of attraction between two masses is proportional to each of the masses, and inversely proportional to the square of the distance between them.

Coulomb's law
The electrostatic force between two charges is proportional to each charge, and inversely proportional to the square of the distance between them.

capacitance
The amount of charge stored per unit p.d.

the farad (F)
The SI unit of capacitance. One farad is one coulomb per volt.

time constant
The product of resistance R and capacitance C in an R–C circuit.

magnetic flux
The magnetic flux passing through an area is the product of the area and the flux density perpendicular to the area.

magnetic flux linkage
The product of the number of turns on a coil and the magnetic flux passing through the coil.

the weber (Wb)	The SI unit of magnetic flux. One weber is one tesla metre squared.
Faraday's law	The e.m.f. induced in a conductor is proportional to the rate at which flux is cut by the conductor.
Lenz's law	An induced current flows in such a direction as to oppose the change that produces it.

Block D4: Thermal physics

| specific heat capacity | The energy required to raise the temperature of 1 kg of a substance by 1 K (or 1°C). |
| specific latent heat | The energy that must be supplied to change the state of 1 kg of a substance, at constant temperature. |

Block D5: Nuclear physics

nucleon number, A	The number of nucleons (protons and neutrons) in a particular nucleus. Also called the **mass number**.
proton number, Z	The number of protons in a particular nucleus. Also called the **atomic number**.
neutron number, N	The number of neutrons in a particular nucleus.
activity	The rate at which nuclei decay in a sample of a radioactive material.
the becquerel (Bq)	The SI unit of activity. One becquerel is one decay per second.
decay constant	The probability that an individual nucleus will decay per unit time.
half-life	The average (or mean) time for the number of nuclei of a nuclide to halve through radioactive decay.

Appendix 4: SI units

The SI system of units is based on seven **fundamental** or **base units**. They are listed in **Table 1** below, together with the quantity of which each is the unit.

You should be familiar with all of these units except the candela.

Most quantities are expressed in **derived units**. For example, area is given in m^2, acceleration in $m\ s^{-2}$. Some derived units are given special names, such as hertz or pascal. Some of these are listed in **Table 2** opposite.

It is often useful to be able to express these derived units in terms of other units. This is shown in the fourth column of the table. The fifth column shows the formulae that relate the corresponding quantities.

Sometimes it is easier to remember the relationship between units, e.g. one volt is one joule per coulomb. At other times it is easier to remember the relationship between quantities, e.g. $F = BIl$. It is a great help if you can translate between quantities and units. Then you need only remember half as many formulae.

Table 3 lists the commonly used **prefixes**, e.g. $1\ \mu F = 1$ microfarad $= 10^{-6}$ F.

Table 1 Fundamental SI units

quantity	unit	abbreviation
mass	kilogram	kg
length	metre	m
time	second	s
temperature	kelvin	K
current	ampere	A
amount of substance	mole	mol
luminous intensity	candela	cd

Table 2 Derived SI units

quantity	unit	abbreviation	in terms of other units	equation
frequency	hertz	Hz	s^{-1}	$f = 1/T$
force	newton	N	$kg\ m\ s^{-2}$	$F = ma$
energy, work	joule	J	N m	$W = Fd$
power	watt	W	$J\ s^{-1}$	$P = W/t$
charge	coulomb	C	A s	$Q = It$
p.d., e.m.f.	volt	V	$J\ C^{-1}$	$W = QV$
resistance	ohm	Ω	$V\ A^{-1}$	$V = IR$
capacitance	farad	F	$C\ V^{-1}$	$Q = CV$
magnetic flux density	tesla	T	$N\ A^{-1}\ m^{-1}$ $Wb\ m^{-2}$	$F = BIl$ $\Phi = AB$
magnetic flux	weber	Wb	V s	$E = d\Phi/dt$
Celsius temperature	degree Celsius	°C	K	$T = \theta + 273$
activity	becquerel	Bq	s^{-1}	$A = dN/dt$

Table 3 Prefixes

factor	prefix	symbol
10^{9}	giga-	G
10^{6}	mega-	M
10^{3}	kilo-	k
10^{-1}	deci-	d
10^{-2}	centi-	c
10^{-3}	milli-	m
10^{-6}	micro-	μ
10^{-9}	nano-	n
10^{-12}	pico-	p

Appendix 5: Electrical circuit symbols

You need to be able to recall and use appropriate circuit symbols; you also need to be able to draw and interpret circuit diagrams that include these symbols.

name of device	symbol
junction of conductors (optional dot)	
conductors crossing (no connection)	
cell	
battery of cells	
open terminals	
indicator or light source	
fixed resistor	
potentiometer (voltage divider)	
light-dependent resistor (LDR)	
thermistor	
ammeter	
voltmeter	
semiconductor diode	
capacitor	
inductor (coil)	

Answers to quick check questions

Block D1: Newton's laws

Energy transfers

1 vectors
2 38.3 m; 766 kJ
3 54 m s^{-1}
4 Work done against air resistance reduces KE.

Momentum

1 mass, KE
2 1.5×10^6 kg m s^{-1} due west
3 Smaller mass × greater velocity = greater mass × smaller velocity.
4 Boy has more momentum, girl has more KE.
5 1 kg m s^{-1}; 2.96 kg m s^{-1}; 3.92 N

Collisions and explosions

1 10 m s^{-1}
2 2 m s^{-1} to right
3 15 m s^{-1}

Newton's laws of motion

1 zero; constant
2 50 kN; GPE increases
3 two contact forces; two forces on person

Block D2: Oscillations and circular motion

Describing circular motion

1 2π; π; $\pi/2$ or 1.57; $\pi/3$ or 1.05; $\pi/4$ or 0.79
2 57.3°; 14.3°; 180°; 360°; 36°
3 524 s; 44 s
4 Velocity, acceleration and force are all changing (vectors).

Centripetal force and acceleration

1 friction with road (and contact force of road, if not horizontal)

2 31.6 N
3 smaller r so larger F
4 1700 m s^{-1}

Simple harmonic motion

1 5 cm; 5.0 s; 0.2 Hz; 1.26 rad s^{-1}
2 0.87 s; 1.15 Hz; 7.2 rad s^{-1}
3 40π rad s^{-1}; 20 Hz
4 at midpoint; at ends of oscillation
5 KE

More about SHM

1 see graphs

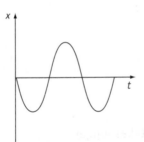

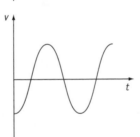

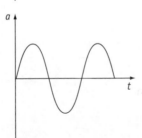

2 4 cm; 0.095 Hz; −3.96 cm
3 $x = 0.2 \sin(\pi t)$; 0.118 m
4 decreases

Block D3: Force fields

Gravitational fields

1 uniform field – parallel field lines; slight separation of field lines; greater separation

2 8 N; 0.8 m

3 533 N

Electric fields

1 parallel, equally spaced lines

2 1.6×10^{-15} N

3 300 V m^{-1} (or 300 N C^{-1})

4 $E \propto V$; $E \propto 1/d$

5 1.92 N

Coulomb's law

1 equal and opposite repulsive forces

2 230 N

3 1×10^{11} N C^{-1}

Capacitors

1 8 μF; 160 μC

2 5

3 40 μF

4 2 μF; 40 mJ

Discharging a capacitor

1 Graph 1 – bigger R, so takes longer.

2 22.3 V

3 240 μA; 32.5 μA

4 both the same

Electromagnetic forces

1 0.66 N; down into paper

2 1.6×10^{-13} N

3 up out of paper

4 0.57 mm

5 the faster; the faster

Electromagnetic induction

1 0.2 T

2 0.079 Wb

3 the first two

4 towards A; positive

Block D4: Thermal physics

Internal energy

1 both: gas

2 Each has half.

3 only heating

4 A: solid; B: solid + liquid; C: liquid; D: liquid + gas; E: gas. Internal energy increasing in all.

Specific heat capacity

1 B

2 18 000 J

3 45.5°C

4 500 J kg^{-1} K^{-1}

Specific latent heat; temperature scales

1 B

2 127.5 kJ

3 2250 kJ kg^{-1}

4 273 K; 373 K; 223 K

5 145 K

Gas equations

1 3.01×10^{24} particles in both

2 241 K (–32°C)

3 4 litres

4 Average KE doubles; stays the same.

Block D5: Nuclear physics

The structure of matter

1 Maximum diffraction when λ is similar to spacing.

2 Momentum increases, wavelength decreases.

3 More chance of 'direct hit', so more back-scattered.

4 5 orders (10^5)

5 10^4

Nuclear structure

1 $^{28}_{14}$Si

2 8 of each

3 A and D; B and C

4 $^{210}_{84}$Po $\rightarrow$ $^{206}_{82}$Pb + $^{4}_{2}$He + γ;

$^{42}_{19}$K $\rightarrow$ $^{42}_{20}$Ca + $^{0}_{-1}$e + γ

Mass–energy conservation

1 1.009 u

2 Mass of neutron > mass of proton + neutron; more mass–energy is needed.

3 4.4×10^9 kg s^{-1}

4 Mass of products < mass of mother.

Radioactive decay

1 γ

2 Background; not all decays detected.

3 1.2×10^{-6} s^{-1}

Radioactive decay equations

1 $A \propto N$, so A decreases as N decreases.

2 $C = C_0 e^{-\lambda t}$

3 34.4 Bq

4 1.23×10^{-4} yr^{-1}

5 115.5 s; 0.30

Answers to
end-of-module questions

1 a product of mass and velocity; a vector quantity

b 1.2×10^5 kg m s^{-1}

c When two or more objects interact, their total momentum remains constant provided no external force acts (i.e. within a closed system).

d The child gains upward momentum, the Earth gains downward momentum. The child's momentum is transferred to the Earth.

2 a Total KE remains constant.

b KE before = $\frac{1}{2} \times 0.005$ kg $\times$ (1 m s^{-1})^2 = KE after.

3 Work done by weight = 6 J; work done by friction = –5 J; work done by normal reaction = 0 J.

4 a 11.76 J

b 10.95 J

c Since energy is conserved, 11.76 J of GPE is transferred to 10.95 J of KE and 0.81 J of frictional heating.

5 a v tangential; F radial

b $F = mv^2/r$

6 a $F = GMm/r^2$; the two forces are equal and opposite.

b 13.5 km s^{-1}

c 3.3×10^5 s

7 a SHM: the acceleration of a mass is directed towards a fixed point and is proportional to its displacement from that point.

b 5 cm; 0.4 s; 2.5 Hz

c $x = 0.05 \sin(5\pi t)$; 0

d 12.3 m s^{-2}; at maximum displacement

8 a direction of field: direction of field line at a point; magnitude: lines closer together = stronger field.

b magnitudes: equal; directions: opposite (Newton's third law)

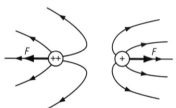

9 a

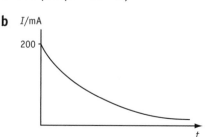

b 20 kV m^{-1}

c 2×10^{-11} N

d Force is unchanged.

10 a 4 mC (4×10^{-3} C)

b 0.8 J

c parallel; 20 µF

d 200 V

e 0.2 J; one quarter

11 a 200 µA (2×10^{-4} A)

b

c 1000 s

d $I = 2 \times 10^{-4} \exp(-t/1000)$

e 120 µA (1.2×10^{-4} A)

12 a

b 0.2 T

c 0.03 N

13 a 3×10^{-4} Wb; 3×10^{-2} Wb

b Flux linking coil is not changing.

c Rotate it, or move it out of field.

14 a A, B, C: internal energy increasing

b A: KE increasing; B: PE increasing; C: KE increasing

c 60 kJ kg^{-1}

15 a the amount of a substance

b $pV = RT$ (because $n = 1$)

c 124 kPa

16 a N = number of molecules; m = mass of one molecule; $<c^2>$ = mean square speed of molecules.

b 6.2×10^{-21} J

17 a alpha-particle scattering

b alpha-particle scattering, high-energy electron scattering

c X-ray diffraction, high-energy electron scattering

18 In fission, a nucleus with large A splits to form two nuclei with smaller A. Total number of nucleons is unchanged. On the graph, fission is represented by movement to left from right-hand end. Binding energy per nucleon increases; energy is released in the process.

19 a 19 protons, 23 neutrons

b $^{42}_{19}\text{K} \rightarrow ^{42}_{20}\text{Ca} + ^{0}_{-1}\text{e} + \text{energy}$

c $\lambda = 0.055$ h^{-1}

20 a $A = 50 \exp(-3 \times 10^{-3}t)$

b 11 Bq

c Not all particles emitted reach Geiger tube; some pass straight through the tube.

Index